How to Build a Cat, from Scratch

By

Cranbrook Kingswood Biology Honors, 2018

Chapter 1
To build a cat from scratch, you will need energy. The energy needed to build a cat comes from sunlight, which must be captured and stored in a useful form.

Chapter 2
Cats don't eat plants, so the trapped energy from sunlight must be transferred to something a cat does eat.

Chapter 3
The captured energy, once it is in the cat, must be stored and then released in a controlled fashion.

Chapter 4
Several different types of building-blocks are required for the construction of the cells of a cat, a mouse, or an apple.

Chapter 5
Once you have harvested and stored the energy needed to construct the cat, and have collected all the required building-blocks, you will need to follow specific instructions to properly assemble these components. The instructions for assembling the building-blocks required in the construction of a cat follow the same rules as those used to assemble an apple tree or a mouse.

Chapter 6
The biological molecules, as well as the activities that utilize, organize and reorganize them occur within

cells. These cells have working subunits, called organelles.

Chapter 7
These cell organelles interact within the working cells. Each cell-type is specialized to do specific work. Like people in a city, each carries out its job, and each depends upon others to carry out their jobs.

Chapter 8
The cells must interact with their environment, which is still inside the cat. Things must be monitored, and certain items allowed into the cells, while certain items must leave. This is constantly regulated, and there are various mechanisms which allow the controlled movement of these items.

Chapter 9
Cells comprise organs, which in turn make up organ systems. The cat itself must be a coordinated, integrated whole, possessing interconnecting systems. This is also true of the apple tree and the mouse.

Chapter 10
Cells comprise organs, which in turn make up organ systems. The cat itself must be a coordinated, integrated whole, possessing interconnecting systems. This is also true of the apple tree and the mouse.

(First, the cat and the mouse...)

(And second, the apple tree...)

Chapter 11
Each cat must be slightly different from all other cats, so some individuals in a population may survive changes in the environment. Therefore, the instructions for building a cat must have a built-in mechanism that allows for variability.

Chapter 12
The cat, the mouse, and the apple tree are all organisms that interact with each other, and with the abiotic environment.

AUTHORS:

How to Build a Cat from Scratch: An instruction manual.

Monosaccharides, Innate Immunity - Rachel Bieler

Disaccharides, Adaptive immunity - Sophie Cronk

Starch, Clonal selection - Madison Fan

Glycogen, 1′, 2′ response to infection - David Hermelin

Protein, Antibodies - Aishwarya Kannan

Fats / oils, T-cells - Saaim Khan

Phospholipids, Thermoregulation - Umaiyal Kogulan

Steroids, Chemical/electrical signals - Cody Krause

DNA structure, Osmoregulation - Shannon Lewand

RNA structure, ADH - Yixuan Li (Olivia)

Cell nucleus, Hormone signaling mechanisms - Hannah Momblanco

Ribosomes, Vertebrate endocrine system - Justine Murdock

Endoplasmic reticulum, Hypothalamus - Dhilan Nagaraju

Golgi, Thyroid - Nicholas Schumacher

Lysosomes, Sex hormones - Hannah Sidberry

Mitochondrion, Pancreatic hormones - Charlotte Trunsky

Chloroplast, Adrenal gland - Gwenyth Woodbury

Cell membranes, Placenta - Paul Yang

Chromosomes - Megan Dixon

Passive transport, Reproductive system of cat - Emma Block

Membrane transport proteins, Fertilization - Shaun Doogal

Osmosis, Sperm formation – Claire Burelle

Active transport, Nervous system overview - Jordan Foxx

Endo/exocytosis, Zygote/blastula/gastrula/organogenesis - Adam French

Breathing/O2 transport, Synapse - Daniel Prakah-Asante

ATP, Neuron structure / Charge across membrane – Jonah Rinaldis

Electron Transport/redox, Nerve impulse - Ian Rosenberg

Glycolysis/pyruvate oxidation, Vertebrate nervous system Chaiya Shah

Citric acid cycle, Brain - Rachel Tang

Metabolic pathways, Sensory receptors - Samuel Wittenberg

Chemiosmosis, Population growth of mice – Yifei Wang (Erik)

Chloroplast structure, Hearing, balance - Kyuhyun Kim (Casey)

ATP and NADPH, Taste/odor - Grant Landry

Light driven reactions, Skeleton - Nicholas Ludwig

Calvin cycle, Cat Eyes - William Mirza

Cell cycle, Muscle - Zijun Zhang (Serena)

Connective tissue, Reticular formation - Gabriel Ervin (Gabe)

Muscle, Angiosperms, dicots (apple) – Andrew Calabrese-Day

Nervous tissue, Basic plant structure - Emily Fernandez

Overview of food processing, Muscle-neuron interaction - Lily Gardella

Mouth/swallow/peristalsis, Aerobic respiration –

Jiayi Hao (Sunny)

Stomach, Primary growth - Hayoung Kim

Small intestine, Secondary growth - Alex Miller

Pancreas /liver, Flower - Jacob Miller

Large intestine/, Pollen/ovules - Khanh Pham

Respiratory system, Seed formation - Lang Qin

Negative pressure/breathing control, Fruit formation (apple) -Lucas Schattenmann

Transport of gases in body, Transpiration - Resha Sheth

Circulation overview, Phloem – Adrian Stone

Heart contraction/SA node, Ethylene - Bozhang Tan (Brody)

Blood vessel structure, Mouse behavior - Grace Theodore

Capillary function, Trophic structure - Jonathan Zhao

Mitosis, muscle - Allison Brook

Cytokinesis, Modifications modifying gene expression - Justin Brown

Growth factors, Cat/mouse population cycles - Marci Edwards

Law of Segregation, Cat/mouse predation – Abby Foltyn

Homologous chromosomes, Herbivory - Zongwei Hu (Tony)

Independent Assortment, Plant hormones - Daniel Juzych

Incomplete dominance, Chromosome structure - Luke Koski

X-linkage: Calico, Control of transcription - Ilina Adhikari (Ilu)

Dominant/Recessive – Manx, RNA splicing - Kelsey Kowal

DNA/RNA structure, Regulation of gene expression – Lauren Philip

DNA replication, Embryonic development – Hyunsoo Ryu

Transcription, signal transduction - Minfei Shen

Translation, nitrogen cycle - Theodore Souris (Theo)

RNA Processing, energy flow/chemical cycling -
Xinyu Sun (Linda)

Ribosome, Energy and trophic structure - Bach Tran

Variation in a species, Carbon cycle - Tommy Wyniemko

Natural selection, Oogenesis - Gefei Zhu (Ken)

How to Build a Cat From Scratch

To build a cat from scratch, you will need energy, and you will need matter. Each of these will take very specific forms. You will also need a set of step-by-step instructions, which have been revised and refined over many years, and you will need mechanisms for following these instructions. This manual will guide you through the building process.

To build a cat from scratch, you will need energy.
The energy needed to build a cat comes from sunlight, which must be captured and stored in a useful form.

Chloroplast*

Energy is essential for life in both plant and animal cells. For plants cells, most of this energy comes from photosynthesis, in which light energy is converted into chemical energy.

Chloroplasts are organelles inside plant and algae cells which carry out photosynthesis, which is the production of organic compounds using light energy. The chloroplast is a very complex organelle which possesses many internal compartments called thylakoids. Chloroplasts have an exceptional history. They are related to small prokaryotes (bacteria) until they were engulfed by a larger cell. The cell ended up forming a symbiotic relationship with the chloroplasts because it provided nourishment and released a large amount of energy. This is known as the endosymbiont theory.

Chloroplast structure*

The chloroplast is located inside the plant cell, and it is the site of photosynthesis. The chloroplast has a double-membrane envelope, meaning it has inner and outer membranes with space in between them. Besides these membranes, there are also internal membrane structures, the thylakoids. Thylakoids are suspended within the stroma, which is the thick fluid. The thylakoids are packed with pigments, the chlorophylls, which trap sunlight to produce food for the plant cell. A stack of the thylakoids is called granum, and the inside of thylakoid is called the thylakoid lumen.

Light Driven Reactions*

Light driven reactions are the first of the two stages of photosynthesis. These reactions happen within the thylakoids, which are membrane-bound chambers located in the chloroplast of a plant. Thylakoids contain a green pigment called chlorophyll, which helps carry out photosynthesis. Light-driven reactions are produced by the intake of light energy and water into the thylakoids. Within the thylakoid membranes, electrons are energized by light energy from the sun. Light driven reactions cause the electrons to flow through a chain of molecules, which causes water's positively charged hydrogen ions to cross an inner membrane, creating a gradient. They eventually join a positively charged NADP causing the synthesis of NADPH, which stores the energized electrons. The energy from the sun is consumed by light reactions, which change it into chemical energy and is stored in NADPH and ATP. When these continuous light driven reactions are completed, waste oxygen gas from the water is released from the thylakoids. Light driven reactions do not create sugar, but the products created in these reactions are used for energy in the second stage of photosynthesis: the Calvin Cycle which creates organic compounds.

Calvin cycle*

The Calvin Cycle is the second stage of photosynthesis and the purpose of it is to produce sugar. It takes place in the stroma (a thick fluid) of the chloroplast. The Calvin Cycle assembles sugars using carbon dioxide and products of light reactions. This is done by way of carbon fixation, the process by which the carbon from CO_2 is attached to a five-carbon compound. The resulting six-carbon compounds breaks in half, into two 3 carbon molecules called 3-PGA. This 3-PGA is converted to G3P (Glyceraldehyde 3-Phosphate). This is achieved due to energy input from NADPH and ATP, where the light's energy has been stored. One of the G3P molecules exits the cycle, and is used to form glucose. At this point, the remaining G3P molecules are used with ATP to regenerate RuBP (Ribulose Bisphosphate), five-carbon the molecule used at the start of the cycle.

ATP and NADPH *

Adenosine Triphosphate, or ATP, is a very important molecule in the process of energy conversion. It is often called "energy currency" because it is traded between different molecules and broken down to liberate the energy it carries. ATP is made up of Adenosine Diphosphate (ADP) and a phosphate group, which are bonded together by the energy originally captured by the plant. The plant converts the light energy from the sun into chemical energy which is stored as ATP. When the plant needs energy, it brings in and breaks down ATP, using its chemical energy to create sugar during a process called the Calvin Cycle.

ATP is functionally like a molecule called NADPH. NADPH is of nicotinamide adenine dinucleotide phosphate, or NADP+, bonded hydrogen. The electrons of the two substances are brought together and bonded using the light energy from the sun. This light energy is converted into chemical energy, and then stored in NADPH until the plant needs it to create sugar. NADPH provides the plant with hydrogen and electrons to reduce the carbon compounds during the Calvin Cycle.

The plant now has glucose, a sugar whose chemical bonds are the stored light energy. The glucose may then be built into branched chains, forming starch.

Cats don't eat plants, so the trapped energy must be transferred to something a cat does eat.

Trophic structure *

Trophic structure refers to the sequence of feeding between organisms on different levels, and is present in every ecosystem. In simple terms, it is the food chain or the food web. The trophic structure starts with the bottom tier, made of autotrophs, also known as producers. All organisms above producers in the trophic structure are called heterotrophs. Most producers get energy from the sun by means of photosynthesis. Producers are usually plants or phytoplankton (microscopic, photosynthesizing organisms). Herbivores (plant-eaters) are primary consumers, which means they eat producers. Subsequent levels are carnivores or insectivores (meat- and insect- eaters), which consume organisms from the levels below themselves. The cat is an example of a secondary consumer, or a primary carnivore, since it eats the mouse, which is a primary consumer, or herbivore. There is also a class of detritivores, which feed off of dead organic material called detritus. Detritus is produced at every level of the trophic structure. Detritivores can either be scavengers (like a crow) or decomposers (like fungi or bacteria). Scavengers feed off of carcasses, while decomposers break down organic materials into inorganic forms. Decomposers are especially vital to the trophic structure, since they convert complex organic materials into a form useful to producers. This way no food supply on any level will be exhausted unless the structure is disrupted.

Herbivory (a mouse eats an apple) *

The presence of herbivores has led to plant adaptations and evolution. When a herbivore eats part of a plant, the latter needs to spend energy on repairing the damage. Therefore, they have developed many defense mechanisms to prevent being eaten, such as thorns or spines or poisons. The herbivores then respond to these new defenses and the two "co-evolve". The mouse, prey of our cat, is an omnivore. Fruits such as apples are one of its favorite kinds of food. Its uses its sense of smell to detect anything bitter in the apple which that might represent a toxic chemical, and eats the good apples, allowing it to grow more juicy and delicious for our cat.

Energy and trophic structure *

Only about 1% of the solar energy striking our planet is actually used in photosynthesis. The energy converted into useful, chemical form is termed the primary production. In an ecosystem, the mass of living organic material, in any form, is referred to as biomass. Different ecosystems have different amounts of primary production.

Energy is lost at each link in the food chain. This phenomenon is demonstrated in the pyramid of production, which shows us the energy loss of energy as material is passed up the food chain. As energy gets passed along the levels, the available amount becomes smaller and smaller, to the point that only fraction of the original input of energy remains. This limits the number of trophic levels an ecosystem can have.

So, a mouse eats an apple, and the cat eats the mouse. This simple food chain transfers the sun's energy to the cat.

The captured energy, once it is in the cat, must be stored and then released in a controlled fashion.

Glycogen *

Glycogen is a polysaccharide made of simple sugar (glucose) molecules. It is stored in the cat's liver and muscles. The glycogen stores the glucose in clumps. The glucose can be released, when the cat needs the sun's stored energy, through the addition of water to the glycogen (hydrolysis). The more glycogen clumps in the muscles the longer the cat can be active without fatigue. Glycogen is made of branched chains of glucose. (Much like starch, in plants, which is how the apple tree stored its glucose.)   Glycogen is a carbohydrate.

The cat stores some of its energy as fat, which acts as a long-term energy storage. A fat is two times more effective than a polysaccharide at storing energy. Fat, a long-term fuel reserve is stored in adipose cells which grow and shrink as the cat adds or extracts fat molecules from them. Fats also serve to cushion vital organs, as well as insulate the cat's body.

ATP *

Adenosine Triphosphate (ATP) provides energy for all mechanical work of the cell. It represents the cell's 'standard currency'. From muscle contraction to the active transport of molecules across a cell's membrane, energy is provided by ATP. ATP is made up of an adenine molecule, a sugar, and three phosphate groups. Energy is released when the bond between the last two phosphates is broken. The energy is transferred in a coupled reaction, so that as ATP loses energy, something else receives it. The resulting molecule (ADP) can have a new phosphate attached, and be reused as energy currency.

ATP and NADPH *

Adenosine Triphosphate, or ATP, is a very important molecule in the process of energy conversion. It is often called "energy currency", because it is traded between different molecules and broken down to free the energy it has stored in one of its bonds. ATP is made up of Adenosine Diphosphate (ADP) and a phosphate group, which are bound together by the reaction between the plant and the light energy. The plant converts the light energy from the sun into chemical energy which is transferred to ATP. When the plant needs energy, it brings in and breaks down ATP, using its chemical energy to create sugar during a process called the Calvin Cycle. Once the energy has been transferred to the cat, the cat, also, will build its own ATP for use in its cells.
ATP is in the same class as a molecule called NADPH. NADPH (nicotinamide adenine dinucleotide phosphate (NADP+), which carries hydrogen and two electrons. The electrons represent the energy from the light of the sun. This energy is used to create sugar.

Breathing/O2 transport *

A cat must be able to harvest energy from stored glucose and other molecules. This process is called cellular respiration. Breathing and cellular respiration work conjointly. When a cat runs down a hallway it inhales, filling its lungs with O2 (Oxygen). The oxygen flows through the bloodstream and into the muscle cells. Inside the cells, cellular respiration occurs generating ATP, which holds readily available energy. However, cellular respiration also produces CO_2 , which the cat exhales and releases into the environment.

4.Redox Reactions *

Carbohydrates are the short-term source of energy for organisms. In order to extract energy from these carbohydrates, a cell must use redox reactions, which represents a transfer of energetic electrons from one molecule to another, arranged in a chain. This is what is called a series of redox reactions, which stands for oxidation-reduction reaction. Oxidation refers to the loss of an electron from a molecule or an atom. Reduction represents the gain of an electron. These two reactions, oxidation and reduction, always happen together because the electrons need somewhere to go. This is how the cells in the cat get energy. This electron transport chain provides energy to transfer many hydrogen ions across a membrane. This uneven distribution of hydrogen powers the formation of ATP.

11.Electron Transport Chain *

For a cell to get energy from carbohydrates, it must use a series of redox reactions. The molecule NAD+ can store two electrons. It strips hydrogens from the carbohydrates. The two electrons and the hydrogen ion bond to NAD+, forming NADH. These electrons then move through a series of redox reactions from the sugars to the oxygen. These reactions occur inside the mitochondria, where the electron carriers go through a set of membrane proteins. Each protein the electrons move through pumps two hydrogen ions to the other side of the membrane. This creates a hydrogen ion gradient. The hydrogen ions then flow back through a protein, ATP-Synthase, which turns and uses that mechanical energy to generate ATP from ADP. This is how the cells of the cat, or any eukaryotic cell, get energy from carbohydrates.

9.Mitochondrion *

Mitochondria are organelles within the cat's cells that transfer the chemical energy consumed in food to ATP (adenosine triphosphate). Surrounding the mitochondria are two phospholipid bilayers, containing various proteins, forming the two membranes. There are two internal parts of the mitochondrion. The first one is a small area located between the two membranes, the intermembrane space. The second, known as the mitochondrial matrix, is located inside of the inner membrane. The mitochondrial matrix contains mitochondrial DNA, ribosomes, and some enzymes. The intermembrane space is covered in cristae, or folds, and is studded with proteins that transfer electrons in the electron transport system. The cristae increase the surface area of the mitochondrial membrane, allowing mitochondria to produce more ATP because there are more places to perform this process. Interestingly, mitochondria they were once prokaryotes (bacterial), until they merged with the eukaryotic cells. This is known as the endosymbiont theory. Due to evolution, the mitochondrion became specialized and are now became an essential part of the eukaryotic cell.

Glycolysis/pyruvate oxidation *

Glycolysis is the first stage of cellular respiration, which is the process to turn sugar useable energy for the cat. The process begins by breaking glucose into two molecules. Glycolysis, the splitting of sugar, is a chemical reaction which breaks down glucose into two pyruvate molecules. There are nine chemical reactions in the process of glycolysis. Steps one through three are reactions in which a six-carbon glucose gains energy from ATP (Adenosine Triphosphate). ATP is formed by the process called substrate level phosphorylation or through chemiosmosis. Two phosphate groups that came off of the ATPs attach to the glucose, forming fructose bis-phosphate. In the fourth step, the six carbon chain splits into 2 three carbon chains with a phosphate group on each. In the fifth step, a redox reaction occurs to create NADH. In steps six through nine of glycolysis, four ATP molecules and two pyruvate molecules are produced. The two pyruvate molecules then go through pyruvate oxidation.

The two pyruvate molecules are prepared for the citric acid cycle. First, the two molecules are transferred from the cytoplasm of the cell to a mitochondrion. The pyruvate molecules don't enter the citric acid cycle itself. Instead, the molecules are oxidized and made into two acetyl coenzyme A (acetyl CoA) molecules. The two acetyl CoA enter into the citric acid cycle.

Citric acid cycle *

The citric acid cycle processes food the cat eats, and harvests energy its body can use. This cycle involves several reactions. (The cycle is also known as the Krebs cycle because of Hans Krebs, a scientist who studied this cycle.) It takes place in the mitochondria of a cell and is the second stage of cellular respiration, following glycolysis. The citric acid cycle breaks down glucose to carbon dioxide.   Acetyl Coenzyme A enters the cycle. Fatty acids, from fats and oils derived from the mouse, can also be broken to this two carbon compound, and enter the cycle. give acetyl CoA and excess protein molecules enter the citric acid cycle too.

Each step of the cycle is helped by enzymes from the mitochondrial matrix. When the two-carbon group enters the cycle, it is joined to a four carbon molecule, making a six carbon group. This six carbon group is called citrate (why this cycle is called the citric acid cycle). The citric acid cycle generates a considerable number of energy-carrying molecules: one FADH2 molecule, three NADH molecules, and one ATP molecule per rotation. Each glucose molecule contributes two acetyl CoA, so two turns of the citric acid cycle process one glucose molecule. During the cycle, six molecules of CO_2 are given off, which the cat exhales.

Chemiosmosis *

Chemiosmosis involves the transfer of hydrogen ions down their concentration gradient, through ATP synthase. The concentration difference between one side of the ATP synthase and the other is created by the electron transport chain. Electrons from various electron carriers travel inside the inner mitochondrial membrane, generating energy for proton channels to pump hydrogen ions out of the mitochondrial matrix into the inter-mitochondrial space, thus creating a concentration gradient of protons. Then, hydrogen ions (protons) move down the concentration gradient from the inter-mitochondrial space back to the mitochondrial matrix through ATP synthase. ATP synthase has the shape of a turbine, so as the hydrogen ions pass through, they turn the turbine, generating energy to catalyze the phosphorylation of ADP into ATP. At the end of the electron transport system, oxygen will come along and pick up two electrons and two protons, forming water.

Aerobic respiration *

The cat uses energy as it performs physical activities. In the cat's muscles, phosphocreatine, a molecule with high energy, transfers a phosphate group to ADP. Aerobic respiration utilizes oxygen to break down glucose for energy. The oxygen, carried to the muscles by the circulatory system, act as the final electron acceptor of the electron transport system. The blood vessels throughout the cat's body transport oxygen and the energy-rich molecule glucose to its muscles. The by-product of the process, carbon dioxide, is eventually carried away from the muscles by the blood. The cat breathes more often while its heart beats faster during exertion, which allows an increase in the amount of oxygen delivered to the muscles. Myoglobin in the cat's muscles also provides stored oxygen for carrying out aerobic respiration. It is a protein which resembles hemoglobin, which is the protein responsible for transporting oxygen in blood. Should the circulatory system be unable to deliver enough oxygen, the cat's muscles will shift to the less efficient anaerobic respiration.

Metabolic pathways *

Metabolic pathways are pathways in a cat, and other organisms, which break down various food molecules, and turn the food into ATP (Adenosine Triphosphate), or create more tissue such as muscle or fat. Metabolic pathways may be catabolic and anabolic pathways. A catabolic pathway gives off energy as it breaks food into smaller molecules, while an anabolic pathway takes up energy as it creates a more complex molecule. An example of a catabolic pathway is cellular respiration. Cellular respiration works by breaking down biochemical nutrients such as a glucose molecule into smaller molecules, transferring the released energy to ATP (Adenosine Triphosphate). The synthesis of proteins from amino acids is considered as an example of an anabolic pathway.

Several different types of building-blocks are required for the construction of the cells of a cat, a mouse, or an apple.

Monosaccharides *

Monosaccharides are the simplest types of sugars. They can have from three to seven carbons, and are defined by containing a single carbonyl group and many hydroxyl groups. The placement of the carbonyl group affects the sugar's classification. If the carbonyl group is on the end of the molecule, it is an aldehyde, but if it is in the middle of the molecule, it is a ketone. All sugars are hydrophilic, (water loving) which means they are polar and dissolve in water.

Glucose is the most common monosaccharide. It is used in organism as an energy source, and can be synthesized into many other compounds, such as starch found in plants and the disaccharide sugars, lactose, found in milk, and sucrose, which is commonly used as table sugar. Complex animals, such as our cat, have a system to keep the amount of glucose in their blood at a constant level (through the antagonistic actions of insulin and glucagon).

Disaccharides *

Disaccharides are formed when two monosaccharides are bonded together by a dehydration reaction. Dehydration causes a covalent bond to be formed between the two monosaccharides. This happens when one water molecule (H_2O) is removed from the original two monosaccharides, taking a hydrogen atom (H) from one monosaccharide and a hydroxyl functional group (-OH) from the other. Some common examples of disaccharides include sucrose, lactose and maltose. Sucrose is formed by the a dehydration reaction between glucose and fructose, and is used as table sugar, and maltose is formed by a dehydration reaction between two glucose molecules. Lactose is formed from glucose and galactose and is a nutrition source for infants during nursing. Enzymes can break disaccharides back down into monosaccharides, which our cat will then use for energy.

Starch *

Starch is the storage polysaccharide of plants. The glucose produced in photosynthesis is linked together into complex chains. Plants withdraw glucose for energy from the starch when energy is needed. Animals (in this case, the mouse), may consume starch as a major part of their diet. Animals possess hydrolytic enzymes capable of breaking down starch by adding water molecules. Glucose is thus be obtained through the breakdown of the starch. To build a cat, glucose consumed from plants would provide the energy necessary to build and maintain the body of the cat.

Fats / oils functions *

Fat is used for long-term energy storage. A fat is twice as effective as a polysaccharide (such as starch or glycogen) for storing energy. In mobile mammals, such as cats, fat is a good molecular alternative to carbohydrates due to its efficiency. Fats also serve to cushion vital organs, as well as to thermally insulate the cat's body. These long-term fuel reserves are stored in adipose cells which swell and shrink as fat is added to or removed from them.

Fat / oil structures *

Fats and oils are two similar classes of lipids that are made up of a 3-carbon glycerol and a long-chain fatty acid. They are both hydrophobic (water fearing), because of their structural makeup of nonpolar bonds between carbon and hydrogen. Another name for a fat is a triglyceride. There are two kinds of fatty acids, based on their structure; saturated and unsaturated. A saturated fatty acid contains a long chain of just carbon and hydrogen bonds. As a result, it has a higher density and is usually solid at room temperature. An unsaturated fatty acid contains a long chain of hydrogens and carbons bonded together, but there is a double bond somewhere in the chain which causes a kink in the fatty acid chain. As a result, it has a lower density (it doesn't pack as tightly) and is therefore usually liquid at room temperature.

Although fats and oils are similar, there is a small, but important difference between them. A fat is solid at room temperature as it mainly contains saturated fatty acids. An oil, on the other hand, contains mainly unsaturated fatty acids and is therefore a liquid at room temperature.

Phospholipids *

Phospholipids are essential to a cat because they make up the bulk of its cellular membranes. Phospholipids are amphipathic molecules, which means they possess both hydrophobic and hydrophilic regions. They consist of a glycerol, phosphate group, and two fatty acids. A glycerol is a three-carbon alcohol with each carbon bearing a hydroxyl group. The two fatty acids are attached to the glycerol and are comprised of a chain of CH2 groups. Phospholipids contain a hydrophilic head and a hydrophobic tail. Phospholipids arrange themselves to form two parallel layers, called a phospholipid bilayer. The hydrophobic tails are oriented toward the interior of the membrane and the hydrophilic heads point outwards toward the cytoplasm or the fluid surrounding the cell. An important property of the phospholipid bilayer is its selective permeability; the ability to only allow certain molecules or ions to pass through, while keeping others from doing so.

Steroids *

Steroids are lipids in which the carbon skeleton contains four bonded rings. Steroids differ from each other based on the different chemical groups attached to the carbon skeleton. Cholesterol is a common steroid in the body of the cat. Steroids may be hormones, (chemical messengers), which are able to pass through the phospholipid bilayer, (a cell's membrane) enter the cell, and bond to receptors inside the cell. Anabolic steroids include the male hormone, testosterone. Testosterone naturally causes a buildup of bone and muscle mass in males, especially around the time of puberty. Female steroid hormones include estrogen and progesterone.

Amino Acids *

When building the cat, you will need to use proteins. The monomers (components) of proteins are amino acids. Proteins may have four levels of organization: primary, secondary, tertiary, and quaternary. The primary structure is the amino acid sequence (a polypeptide). The sequence is determined by instructions in the genes. The secondary structure includes the hydrogen bonds, allowing for certain shapes. The polypeptide chain can either form a helical coil or a pleated sheet. Tertiary structure involves the interactions between the "R" (variable) groups and determine the overall shape of the protein. The tertiary structure is determined by 4 main factors; hydrogen bonds with R groups of amino acid subunits, ionic attraction between R groups with positive and negative charges, hydrophobic interactions of R groups with water, and disulfide bridges, which link the sulfur atoms of 2 cysteine amino acids, causing the two cysteines sulfhydryl groups to bond. Quaternary protein structure is based on the interactions among two or more complete polypeptides.

Protein *

Proteins are macromolecules composed of amino acids. They have many functions and are the most versatile components of cells. Some examples of different proteins are enzymes, structural proteins, storage proteins, transport proteins, regulatory proteins, motile proteins, and protective proteins. Proteins are divided into two structural catagories - globular and fibrous. Globular proteins, such as enzymes, form a complex clump. Fibrous proteins such as proteins making up hair, tendons, and ligaments are mainly used for structure and are long and thin. A protein's function is dependent on its shape. They can sometimes lose function in a process called denaturation, in which their shape is modified. This may result from a change in the surrounding pH or temperature.

Once you have harvested and stored the energy needed to construct the cat, and have collected all the required building-blocks, you will need to follow specific instructions to properly assemble these components. These instructions follow the same rules as those used to assemble an apple tree or a mouse.

DNA/RNA comparative structure *

DNA and RNA are nucleic acids which that have long chains of chemical subunits, called nucleotides. Each nucleotide contains a nitrogenous, ringed base, a sugar, (ribose for RNA, and deoxyribose for DNA), and a phosphate group. The nucleotides are connected by covalent bonds between the sugars and the phosphates, forming a sugar-phosphate backbone. DNA is composed of two polynucleotide chains twisted together to form a double helix. The four bases forming the ladder-like rungs adenine, thymine, guanine, and cytosine. In DNA, the ribose is missing an OH group so it is called deoxyribose. RNA is single stranded and its bases are cytosine, adenine, guanine, and uracil. RNA has ribose in its nucleotides.

DNA structure *

Deoxyribonucleic acid, or DNA, is arranged in a structure called the double helix, two polynucleotides twisted around each other to make a shape like a twisted ladder. Each nucleotide consists of a phosphate group, a sugar, and a nitrogenous base. In the case of DNA, the sugar would be deoxyribose and the nitrogenous base would be one of four options: adenine, thymine, cytosine, or guanine. Bases pair by means of multiple hydrogen bonds, which hold the DNA in its double helix structure. Adenine and guanine are part of a group called purines. They are double-ring structures and cannot pair with each other. Thymine and cytosine each have a single-ring structure and are part of the pyrimidine group. They, too, cannot pair with each other. The correct pairing of bases is adenine to thymine and guanine to cytosine. When adenine is bonded to thymine, it forms two hydrogen bonds, and when guanine bonds with cytosine, three hydrogen bonds are created. By themselves, hydrogen bonds are very weak, but when placed with all the other hydrogen bonds they make a very stable structure.

RNA structure *

RNA is typically a single stranded ribonucleic acid, which means its structure is only one single chain, and ribonucleotides are its monomers. Ribonucleotides are linked by phosphodiester bonds, which join successive sugar and phosphate groups of the nucleotide monomers. A ribonucleotide in the RNA chain consists of the pentose sugar ribose (a pentose is a monosaccharide which contains five carbon atoms), a phosphate group, and one of four nitrogenous bases (adenine, cytosine, guanine, uracil).

Ribosomes *

The ribosomes are essential parts of cells that will form your cat. Ribosomes are formed of two subunits: a large subunit and a small subunit. Both of these subunits consist of proteins and rRNA (ribosomal RNA). The rRNA, created in the nucleolus, and the proteins, assemble to form the subunits, which then exit into the cytoplasm separately. Soon after the subunits exit the cytoplasm, they come together and form a functioning ribosome. These ribosomes vary in number due to how active the cell is in the process of making proteins; the more active the cell is, the more ribosomes there are in the cell. Ribosomes also vary in their location within the cell. They can be found freely suspended in the cell's cytosol or bound to the Rough Endoplasmic Reticulum.

Ribosomes are essential for the creation of proteins in the cat's cells. Most proteins are created on the ribosomes found in the cytosol, such as the enzymes which catalyze breaking down a sugar. However, the bound ribosomes create a number of proteins with a variety of functions. Bound ribosomes create proteins which are inserted into the cell's membrane, sent to other part of the cell, or are secreted from the cell. Next is how a ribosome specifically creates a protein for your cat's cells to use.

Polypeptides are made by a ribosome through a number of steps. First, mRNA fits into the binding site of the small subunit of the ribosome and tRNA binds at one of the two tRNA binding sites on the large subunit of the ribosome. Once bound, the translation begins. An initiator tRNA binds specifically to the start codons of mRNA to start protein synthesis. This initiator binds to the P site on the large subunit of the ribosome; this is where the peptide, as it is being synthesized, will be held. To continue the polypeptide being made, a new tRNA comes to the ribosome and binds to the A site, bearing a new amino acid. Once the codons of the mRNA at the ribosome recognize the anticodons on the tRNA, it allows a peptide bond to form between the peptide chain and the new amino acid. The peptide bond allows the peptide, now connected to the amino acid, which is connected to the tRNA at the A site, to let the tRNA at the P site go. Now only the tRNA at the A site remains. This tRNA and polypeptide chain move to the P site and this process continues in a cycle until a stop codon of the mRNA reaches the A site of the ribosome, either UAA, UAG, or UGA. Ribosomes and their functions are essential to the building of your cat.

Chromosomes *

Chromosomes play a key role in the makeup of a housecat. They carry genetic information stored in tightly coiled DNA. The cells of the domestic cat contains 38 chromosomes, (compared to 46 in humans). Chromosomes come in pairs - one from the mother, and one from the father, and a cat has 19 pairs of chromosomes. These pairs are called homologous chromosomes, meaning they have the ability to carry different genetic information for the same characteristic – allowing offspring to be differentiated from one another. During reproduction, the chromosome pairs each duplicate into parts called sister chromatids. Then, they separate completely so the sister chromatids can also separate into four different haploid cells. This process is called meiosis, and the resulting daughter cells will each have 19 chromosomes. Therefore, when an egg is fertilized by the sperm, it will regain the remaining 19 chromosomes, making a sum of 38 chromosomes.

Genome *

The genome is the complete set of chromosomes possessed by an individual. To build a housecat we need 38 chromosomes. A chromosome is made of a strand of DNA and many proteins. Chromosomes are where the DNA is kept and organized into genes. All the chromosomes exist in pairs, with one of each pair coming from the mother and the other coming from the father. Most of these chromosome pairs are homologous, meaning that the two chromosomes are identical in size, color and all characteristics, though they carry different forms of any given gene. One pair of chromosomes, however, known as sex chromosomes, determines the sex of the kitten. This pair is either an XX or an XY. Females possess two X chromosomes and males possess one X and one Y. The sex of the kitten is determined by what chromosome the male passes down. If a Y-bearing sperm cell fertilizes an egg, it will result in a male kitten. If, instead, an X-bearing sperm cell fertilizes the egg, a female kitten will result.

Chromosome structure and gene expression *

Every organism must be able to turn certain genes on and off in response to signals from its internal and external environment. Similarly, all multicellular organisms must undergo a process called differentiation, in which their cells become specialized. This process occurs during the maturation of the organism from embryo to adult. Every cell must then maintain its specific structure and function, in which only some of its available genes are expressed. Therefore, cell-type differences are due to selective gene expression, rather than different genes being available.

To form chromosomes, a double helix of DNA is wrapped around small proteins called histones, creating a DNA-histone complex. This creates an appearance of beads on a string; each "bead" consists of DNA tightly wound around 8 histone protein molecules and is called a nucleosome. Next, the beaded string is packed tightly into a tight helical fiber, which coils further into a thick "supercoil", the actual body of the chromosome. Further looping, folding, and coiling within the chromosome further compacts the string. As you can tell, the chromosomes of your cat are extremely compact, as there is a lot of genetic material compressed into very little space.

This DNA packaging to form the structure of chromosomes can be an efficient method of selective gene expression, as this tight packaging blocks certain genes from expressing themselves in certain cells of your cat.

Cell cycle *

The cell cycle is an ordered set of events involving cell growth (interphase), and cell division (mitosis) of that cell into two daughter cells. It is mainly composed of 2 stages: interphase and mitotic phase. Cats, mice and apple trees rely on this to grow and survive.

Interphase is a long interval of the cell cycle (about 90 percent of the entire cell cycle). During this phase, the cell replicates its DNA and performs its functions. There are 3 sub-phases of interphase: the G1 phase ("first gap"), the S phase ("synthesis" of DNA or DNA replication) and the G2 phase. During the G_1 phase, the cell grows. During the S phase, chromosomes are duplicated via DNA replication. During the G_2 phase, the cell continues to grow and gets ready for cell division.

The second phase is the mitotic phase, which is only 10 percent of the entire cell cycle. During this phase, the cell division takes place as a single cell divides into two identical daughter cells. There are 2 concurring stages of the mitotic phase: mitosis and cytokinesis. Mitosis is a nuclear division, which involves the separation of identical (sister) chromosomes. Cytokinesis happens right before the end of mitosis. During cytokinesis, the cytoplasm divides into two daughter cells. The whole mitotic phase brings about the the creation of two identical daughter cells, each with a nucleus, a plasma membrane and different organelles surrounded by cytoplasm. Cats are eukaryotes, (they have membrane-bound organelles). For eukaryotes, mitosis or cell cycle mainly allows growth, repair and replacement of cells.

Mitosis *

Mitosis is the process within the cell cycle wherein duplicated chromosomes separate into two daughter nuclei, each with the same DNA. The parent cell gives one copy of every chromosome to each of its two daughter cells. Therefore, the number of chromosomes in the nuclei stays the same through consecutive mitotic divisions. During interphase, the cell carrys out normal activities, while preparing for mitosis. There are four phases in the process of mitosis: Prophase, Metaphase, Anaphase, and Telophase. In early Prophase, the nucleolus and nuclear envelope begin to disappear. In late Prophase, the chromosomes continue to thicken and shorten, and the centrioles move to the opposite poles of the cell. During Metaphase, the chromosomes line up along the cell's mid-plate. During Anaphase, the separation begins. Each sister chromosome moves to opposite sides of the cell. In Telophase, the chromosomes are surrounded by newly forming nuclear envelopes, and cytokinesis occurs, resulting in two separate cells, each with the same DNA.

Cytokinesis *

The final stage of the mitotic phase in a cell is called cytokinesis and usually occurs during the time of telophase. During this phase, the parent cell divides into two separate daughter cells. Though the cells are similar, cytokinesis differs between plants and animals. A cat has only animal cells inside it therefore we should focus on the division of animal cells.

Animal cells can divide in a complex manner which includes many steps. The first sign of this splitting of the parent cell is a small groove forming around the cell. This is called a cleavage furrow. A ring made of actin microfilaments forms around this furrow. The two microfilaments react with each other causing the ring to become smaller and smaller. The ring contracts, deepening the cleavage in the cell until the parent cell splits into two daughter cells.

Telophase of mitosis occurs at the same time as cytokinesis, and has a dual role. Telophase consists of the pulling apart of the newly duplicated chromosomes into the two new daughter cells. Once the cell starts dividing, chromosomes should be pulled apart from each other, into each of their new daughter cells. A problem can occur if a chromosome can't get through the dividing cell, giving one daughter cell an extra chromosome, while the other is missing a chromosome. This can cause many problems in these two daughter cells.

Growth factors *

Growth factors are polypeptides which stimulate cell division and regulate cell survival. They act as signals to tell the cell to begin going through the cell cycle and divide. The growth factors deplete quickly, limiting production. There are at least 50 different kinds of growth factors and may be proteins, hormones, or vitamins. They also have their own families and each family has a different use. For example, one family is the Platelet-derived Growth Factor Family, PDGF. They are a major protein growth factor. Different cell types have their own reaction to different growth factors.

Law of Segregation *

The Law of Segregation is Gregor Mendel's first law. The central premises of this law are: (1) traits don't blend in heterozygotes. Individuals who are heterozygous for a certain gene carry two different alleles (two different forms of a given gene). (2) During meiosis, alleles segregate from each other. A heterozygous condition is a condition in which a gene contains different alleles for a given characteristic. (3) Each gamete (reproductive cell) has an equal chance of receiving either member of a pair of alleles. Each egg or sperm cell carries just one allele for each trait. Once an egg and a sperm fuse, the resulting zygote possesses both forms of the gene. In the

heterozygote, one may be expressed, while the other may be masked.
Homologous chromosomes *

An important factor when building your cat is knowing how homologous chromosomes increase its genetic variations. Two similar chromosomes, known as homologous chromosomes or homologs, are sorted into a pair during meiosis, a type of cell division that results in egg or sperm. They are identical in centromere position and length. One from the female parent and one from the male, the chromosomes carry alleles, forms of each gene, at each of their loci. However, the alleles themselves do not have to be identical; they can be either homozygous, if both are dominant or recessive, or heterozygous, if one is dominant and the other is recessive. When homologs cross over during meiosis, two pairs have their chromatids attached at sites called the chiasmata. They exchange segments of their genetic material and produce new combinations of genes, therefore becoming more genetically varied.

Independent Assortment *

The law of independent assortment is a principle outlined by Gregor Mendel, stating that when two characteristics are inherited, the inheritance of one character has no impact on the inheritance of another. In other words, the law of independent assortment states the alleles of two different genes get sorted into gametes independently of each other, if they're on different chromosomes. The law of independent assortment is also commonly referred to as Mendel's second law. Mendel formulated this principle after performing dihybrid crosses between plants in which two traits, such as seed color and pod color, differed from one another. After these plants were allowed to self-pollinate, he noticed the same ratio of 9:3:3:1 appeared. Mendel concluded traits are transferred to the offspring independently. For example a plant with the dominant traits of green pod color (RR) and yellow seed color (YY) being cross pollinated with a plant with yellow pod color (rr) and green seeds (yy). If the offspring are allowed to self-pollinate, a 9:3:3:1 ratio will be seen in the next generation. About nine plants will have green pods and yellow seeds, three will have green pods and green seeds, three will have yellow pods and yellow seeds and one will have a yellow pod and green seeds. When building a cat from scratch the same 9:3:3:1 ratio would occur if both parents were heterozygous for each of two traits, and those genes were located on different chromosomes.

Incomplete dominance *

An important factor when building your cat from scratch would be its alleles and their level of dominance in a genome. Incomplete dominance is a form of intermediate inheritance, in which two parent alleles, or traits, mix. One of the parent alleles has some dominance over the other, but not complete dominance. A usual phenotypic ratio of offspring in situations of complete dominance of one allele over another would be 3:1, where the dominant allele has three times the possibility of appearing in the offspring as the recessive allele. In contrast, in cases of incomplete dominance, the phenotypic ratio is 1:2:1, where the dominant and recessive alleles each have a 25% chance of appearing in the offspring, and a blended mixture of the two has a 50% chance. Thus, neither allele has complete dominance over the other. An example of incomplete dominance when building a cat would be the mating of a white purebred cat and a brown purebred cat. The white color allele has no complete dominance over the brown color allele, and the brown has no complete dominance over the white. Instead, some forms of each of these genes are expressed, making an interesting combination of the two colors, an orange cat.

X-linkage: Calico *

A calico cat's color is based on the alleles (or a form of a gene) located on the X chromosome. How female calicos get their color is different than how males get theirs. For males, they only have 3 colors: white, black, and red-orange. These colors are decided by the alleles on the X chromosomes they receive. If they get an X chromosome with an allele for orange fur, the males have no choice but to show this color of fur in that one spot on their body. Because of this, males don't display the calico markings. For females, they have a mixture of white, black, and red-orange. This is because they receive 2 X chromosomes from their parents and the mixture of colors is when they get, for example, an X chromosome with an allele for black fur and an X chromosome with an allele for

orange fur. This is why the females have many colors on their fur.
Dominant/Recessive – Manx *

Genes control a cat's traits. There are usually two types of genes; dominant and recessive. Dominant traits are heritable characteristics which are controlled by genes and are presented in the offspring when passed down by a parent. For a dominant trait to occur, the trait only has to be passed down by one parent. Recessive traits are characteristics passed down by both parents. For a recessive trait to be expressed, it must be passed down by both parents and cannot be masked by a dominant gene. Let us examine the dominant trait Manx for example. This trait is heterozygous for dominant mutation. This means there are two different alleles, or alternatives, for a certain gene trait. When building a manx cat, they must have one dominant M and one normal recessive m gene. This results in the absence of a tail or a short tail (manx phenotype). Should a cat receive two of these dominant genes, its skeleton would be so deformed that it not survive to birth.

Protein assemblies control transcription *
The transcription initiation is very important. It is controlled by the complex protein assemblies. It is composed of regulatory proteins called activators and repressors. They both bind to DNA and tell the RNA what to turn on or off in the genes. The activators are more important than the repressors in eukaryotic cells. This is because the default genes for eukaryotic

cells are off except for the most frequently used genes such as the ones that code for the enzymes responsible for Glycolysis. For these eukaryotic cells to function, transcription factors are needed. There are 4 steps to Gene Transcription Initiation. First, the activator proteins bind to the enhancer, which are DNA control sequences. Then, proteins that bend the DNA pull the activators closer to the promoter, which is the binding site for RNA polymerase. The bound activators and transcription factor proteins interact and bond to the promoter, which facilitates the initiation of transcription. The last step is when the polymerase moves down gene strand to produce an RNA strand. This is important in cats because it controls what genes need to be turned on or off.

RNA splicing *

The process of splicing may assist in controlling the flow of the mRNA from the nucleus to the cytoplasm. RNA is attached to the molecules of splicing machinery until splicing is completed, rendering it unable to pass through the nuclear pores. Sometimes a cell can carry out splicing in more than one way. This causes different mRNA molecules to be made by the same transcript of RNA. For example, one mRNA molecule may end up with one color exon, a segment of RNA which contains information to build a peptide sequence or protein, whereas the other mRNA molecule may have a completely different color. With this type of RNA splicing, more than one type of polypeptide can be produced from one single gene.

Multiple mechanisms regulate gene expression *

The steps involved in gene regulation are much like water in pipes. Valves can control the flow of water at key points. Genetic information flows from a chromosome through many control points (like the valves) to create an active protein in the cytoplasm of the cell. Chromosomes contain genes (DNA). The control points can turn a process on or off and can speed or slow the process. Only a few of these control points may be important for any particular protein. A very important control point for gene expression is for the start of transcription. This occurs in the nucleus. The processing of RNA in the nucleus adds nucleotides to the cap and tail of the mature RNA and removes introns. Once mRNA enters the cytoplasm, mRNA translation and breakdown can be regulated. Following the translation stage, the new polypeptide may need to be activated. Eventually, all proteins are broken down into amino acids. Control of transcription, control of mRNA breakdown, control of translation, control of protein activation, and control of protein breakdown can operate in a cat's cell.

DNA Replication *

DNA replication is a process of copying DNA. It starts with the separation of a parent DNA into two strands. Each strand of the parent DNA attracts free nucleotides within the nucleus (Guanine pairs with

Cytosine, and Adenine pairs with Thymine.) In this process enzymes, called DNA polymerases, link nucleotides to form new daughter strands. Each of those daughter DNAs has one parental strand and one daughter strand. DNA replication begins at origins of replication. When a protein attaches to the origin of replication, it initiates DNA replication. Replication proceeds in both directions, forming replication bubbles. Free nucleotides attach to each strand to form daughter strands from 5'end to 3'end direction only. Therefore, one of the two daughter strands can be formed as one continuous strand. However, to make the other daughter strand, DNA polymerases must work backward from the forking point, synthesizing many short pieces of daughter strands called Okazaki fragments as the parent strand opens. Then, another enzyme called DNA ligase, links those pieces to make a single DNA strand. The purpose of DNA replication is to duplicate the cell's DNA during cell division. Replication is required to form a kitten from an embryo, a cat from a kitten, and new cells in the grown cat.

Transcription *

To complete gene expression, the first step is to complete the transcription.
Transcription is how the instructions for protein synthesis, stored in DNA, is put into use. During transcription, RNA is synthesized under the direction of DNA. The initiation of transcription is important to regulating gene expression. You need to use regulatory proteins, activators and repressors, to

control the initiation of the transcription of genes. Usually, any one cell only needs to transcribe a small percentage of its genes in order to carry out its functions.

Translation *

After RNA is transcribed, it is sent to a ribosome, either one attached to the Rough ER or free-floating in the cytoplasm. Once the RNA reaches its destination, it is fed through the ribosome and its nucleotides are "read" in blocks of three called codons. Out of the four nucleotides, a total of sixty-four possible codons can be registered. Since only twenty different amino acids can be synthesized, all amino acids except for methionine and tryptophan can be read from multiple codons, though these codons are usually similar. For example, UAU and UAC both translate to Tyrosine. Certain codons (and the amino acids which they represent) will denote the start or end of an RNA sequence (as well as the resulting polypeptide chain). After having received its instructions, the ribosome makes tRNA, or transfer RNA, which consists of the complements of the RNA that the ribosome read, as well as being attached to the corresponding amino acid. For example, if a UAC is read, the ribosome spits out an AUG attached to a tyrosine. When the next tRNA is paired, the amino from the previous codon bonds to the one from the new codon, forming a polypeptide chain (PPC) that runs along the length of the RNA. Once the entire strand of RNA has been completely paired up with anticodons, and therefore appropriate amino acids, it is re-read multiple times, and is eventually degraded. Meanwhile the PPC, which is currently in the primary, or first, stage of protein structure, is shipped off to the Golgi Apparatus to be refined, reshaped, and combined with other PPCs by multiple enzymes to form a finished product which, in turn, leaves the trans face in a vesicle and is sent off to wherever it is needed. This whole process automatic, so as long as you have provided the proper ingredients you don't have to fret the details of this one.

RNA Processing *

RNA, known scientifically ribonucleic acid acts as a supporting nucleic acid for DNA (Deoxyribonucleic Acid). RNA processing means for RNA to transfer the information from the DNA and send it to the area where the information from the DNA would be translated into an amino acid sequence. The most common of the three kinds of RNA to be processed is mRNA, the messenger RNA. For most eukaryotic genes, there are regions in the RNA which are coding, called exons. The exons are separated by noncoding regions, introns. During the transcription, both exons and introns from the RNA are copied. After they are copied, they go through a process called RNA splicing which means the introns, the non-coding regions are discarded and exons, coding regions, are joined to form a continuous sequence of codes. When all of the above steps are done, the fully developed mRNA will wait for translation.

During RNA processing, adenines are added to one end of the transcript, and a guanosine cap is added to the other.

Modifications of gene expression *

To modify a chromosome's function without directly affecting the gene you can add a diverse variety of chemicals. You will want to regulate the gene expression of the chromosome to make it do what you want. To contract the chromosome you will need to add a methyl group. This will reduce transcription of the chromosome. Adding acetyl to some chromosome-bound amino acids will do the opposite opening up the chromosome, causing more transcription to occur. Not only chromosomes can be modified chemically but DNA can also be changed. The addition of extra methyl to the DNA bases, (methylation), will turn a gene off. If you decide to remove the extra methyl(s), the DNA will turn back on performing the same function it did before. Methylation will usually occur on the base cytosine in the DNA, not changing the DNA directly but making it turned on or turned off. DNA modification can be maintained throughout the creation of new DNA. This process of inheriting a trait without modifying a gene is called epigenetic inheritance. This process of chemical modification to a base on DNA can be reversed if the function of the modification is no longer needed even after the DNA has been transmitted between parent and daughter cells.

The biological molecules, as well as the activities that utilize, organize and reorganize them occur within cells. These cells have working subunits, called organelles.

Cell membranes *

What is the cell membrane, and why is it so important? The cell membrane is the outer layer of the cell and is made mostly out of phospholipids. It is selectively permeable, which means it controls what goes in and out of the cell. Diffusion is the tendency of the particles of any substance to spread out into the available space. Some small, nonpolar substances pass through the cell membrane via diffusion. This occurs because the random movement of particles causes them to even out on both sides. Therefore, if the concentration of a substance is greater on the inside of a cell than on the outside of the cell, the substance will pass through the membrane until both sides of the cell have an equal concentration of the substance. A special form of diffusion is called osmosis. This refers to the net movement of water through a selectively permeable membrane. Although water passes freely through the cell membrane, osmosis moves it from an area of higher concentration to an area of lower concentration. Other larger, polar substances move through the membrane by means of membrane proteins. It is active transport, and requires the input of energy.

Ribosome structure / function *

Among the components which that carry out translation for a cell, ribosomes are the final ones. A ribosome is made of one large and one small subunits, each consists of proteins and ribosomal RNA (rRNA). The large subunit possesses a tRNA binding site, and there is a mRNA binding site located on the small subunit. Both of the subunits contribute to the formation of polypeptide chains. The tRNA, carrying the sequence of bases, carries a specific amino to a growing polypeptide chain.

tRNA possesses anti anticodon, to match the codon on the mRNA, allowing it top place the amino acid it is carrying in the appropriate place on the growing chain.

Cell nucleus *

The cell nucleus, where the DNA is housed, is the site in which the cat stores information to make proteins. The nucleus has the cell's genetic "blueprint", coded in DNA, which directs protein synthesis. The nucleus holds many proteins and DNA organized into chromosomes. The nucleus has its own double membrane called the nuclear envelope, made from two phospholipid bilayers. The endoplasmic reticulum is a continuation of the nuclear envelope. The nuclear membrane separates the nucleus from the other parts of the cell. Embedded in the nuclear envelope are protein lined pores which control what moves in and out of the nucleus. Ribosomal RNA (rRNA) is brought in through the pores and is synthesized in the nucleolus. There can be many nucleoli in the nucleus based on the activity of the cell. The proteins from the cytoplasm are synthesized with the rRNA to create the subunits of ribosomes from the DNA "blueprint". After the proteins are synthesized, they leave the nucleus through the pores to form functional ribosomes.

Endoplasmic Reticulum *

The endoplasmic reticulum is a fundamental part of the endomembrane system. In secretory cells, it processes the messenger RNA and is essential in the release of the resultant protein from the cell. These proteins are stored in vesicles that then travel to the Golgi complex and are released through the cell membrane. Two types of endoplasmic reticulum (ER), are the smooth and rough ER, which describes their surfaces.

The Rough ER is characterized by the ribosomes attached to the outer surface. Messenger RNA is fed through the ribosomes and polypeptides are created. Secretory proteins are made and transported throughout the cell in vesicles. Glycoproteins can form from polypeptides which are linked to polysaccharide chains.

The smooth ER stores calcium ions involved in pumping calcium ions into the cytoplasm. The Smooth ER is classified as having no ribosomes on the outer surface. They are found in liver cells and are known for lipid, phospholipid, and steroid synthesis. Cell detoxification, another function of the smooth ER, breaks down certain chemicals in the liver (ie.

Alcohol, Tylenol, as well as some things more likely to be ingested by a cat). The endoplasmic reticulum processes the messenger RNA and finally releases its secretory proteins stored in vesicles. The transport vesicles then travel to the Golgi complex where lysosomes and other types of vesicles are made.

27.Golgi *

The Golgi apparatus is a structure of many thin pouches. It is responsible for storing, finishing, and sending cell products that it receives from vesicles that bud off the rough ER. It fuses many of these ER transport vesicles together to form its first pouch. The Golgi apparatus' pouches are not connected together so the Golgi enzymes must travel from pouch to pouch to finish each cell product and provide them with tags they will need to be sorted for transport out of the Golgi apparatus. The last sac sorts the finished cell products by the tags they have been given in the other sacs, then splits off from the Golgi into transport vesicles, each vesicle containing finished cell products set for the same destination.

Lysosomes *

The lysosome is involved in a few vital processes in the body. One of its main functions is to be in the digestive process. In this process, the lysosome has the function of working with the food vacuoles, formed through an engulfing process carried out by the cat's white blood cells. It fuses with the food vesicle and dumps its digestive enzymes into the resulting vesicle.

These cell organelles interact within the working cells. Each cell-type is specialized to do specific work. Like people in a city, each carries out its job, and each depends upon others to carry out their jobs.

Taste/odor *

The senses of smell and taste are controlled by
sensory receptors, which are a combination of
neurons and other cells which read signals from the
outside world and internal organs. Taste and smell
are two very similar senses, but they do have their
differences. The process of tasting something works
in a series of steps. First, the molecules of the
ingested substance travel to the taste buds. Taste
buds are the sensory receptor cells for tasting. There
are five different tastes: sweet, sour, bitter, salty, and
umami. Umami is the Japanese word for delicious,
and represents the flavor brought out by cheeses or
meats. It is activated by glutamate, which is an
amino acid. In the taste buds, the ingested chemicals
bind to membrane embedded proteins called sweet
receptors. This process of binding causes a signal
transduction pathway, which causes ion channels in
the membrane to open. The stronger the stimulus
(taste/flavor), the higher the receptor potential. This
receptor potential sends signals into the central
nervous system. Back in the taste bud, every receptor
cell forms a synapse with an individual sensory
neuron. The receptor cell then lets out
neurotransmitters, which cause a flow of action
potentials in the sensory neuron of the cell. Those
action potentials are what cause the cat to taste what
it has ingested.

Chemoreceptors are cells that allow the cat to sense
taste and smell. Airborne chemicals are detected by
chemoreceptors in the nose. Those chemicals are
then dissolved into the cat's mucus, which covers the

entire nasal cavity. Olfactory chemoreceptors are located in the upper nasal cavity. They send impulses through axons into the olfactory bulb, which is located in the brain. Olfactory receptors have cilia which stick into the mucus in the nasal cavity. When a chemical is diffused into the mucus, these cilia can bind with this chemical. This bond causes membrane depolarization and creates action potentials.

Connective tissue *

Connective tissue is an absolute necessity to any animal, and includes the following: blood, bones, and cartilage. The cells which form the connective tissue are responsible for the production and secretion of the matrix, which is often made up of a network of fibers rooted in a lipid, jelly or even a solid. These connective tissues fall into six large groupings.

Loose connective tissue is the most widespread of the connective tissues as it is comprised of a series of loosely connected fibers in a watery fluid. A diverse amount of the fibers in this system are composed of a strong substance known as collagen. Other fibers are rubbery, making their tissue durable. Loose tissue has one main job: to bind the epithelia to the other tissues which that holds an organ in place.

Fibrous connective tissues have a matrix of tightly packed collagens. This organization of collagens makes the durability high, which is a very good thing considering this tissue is responsible for the joints and tendons which allow the bones and the muscles to work together.

Adipose tissue has the job of maintaining the fat. It does it through large cells which that are tightly packed. The matrix of these tissues is very sparse with loose fibers and fluid. The job of this tissue is to use the fat as barriers and cushions. Inside the cell there is a pocket which swells and shrinks as fat is deposited or taken away from it.

Cartilage is responsible for forming a strong material which helps with the joints. Once again we see the appearance of collagen as its matrix consists of collagen fibers rooted into an elastic material. Cartilage is fixed to the bone to absorb shock, providing a cushioning surface for a disk. It also serves the purpose of supporting the ears and the nose.

Bones are made up of collagen fibers, but unlike the other connective tissues these collagen fibers are rooted in a hard calcium compound, which also includes magnesium and phosphate. The way the minerals are combined makes sure that the bone is hard without being brittle. The tight parts of the bone have circular canals containing blood vessels. Contrary to popular belief the cat's bones are just as alive as the rest of the cat.

Blood's main job is to transport itself along with other substances throughout the body. It has a vast matrix of a liquid plasma that transports red and white blood cells as well as platelets, which help to protect the cat from infection.

Bones allow the cat to keep its structure, the blood transfers essential nutrients and protects the body from infection. Ligaments, tendons and collagen are important for maintaining the position of structures.

Muscle cells*

The cat's muscle tissue helps the body move and control the heart to distribute blood. To build muscles, the cat must first create the cells, also known as muscle fibers, which make up muscles. These fibers contain proteins that allow contractions to take place in the muscle, whether they are voluntary or involuntary. Three types of muscle fibers comprise a cat's muscular system: skeletal muscle, cardiac muscle, and smooth muscle.

Skeletal muscles cause the voluntary motions of the body. When building these fibers, arrange the fibers in a striped pattern, called striations. Be sure to use tendons to connect to the bones nearby. Without tendons, the muscle cannot move the bone, not allowing them to do their job.

Cardiac muscle is the contractile muscle tissue of the heart. This muscle is involuntary, meaning the nervous system controls the movement of the tissue. Be sure to also striate the muscle. Also make sure cardiac muscle is branched, meeting at various intersections to pass on the message to contract, allowing the heart to beat.

Smooth muscle is found in the walls of organs such as the intestines, the digestive tract, and arteries. It helps control blood pressure and the distribution of blood. Smooth muscle is involuntary and should be arranged in a spindle shape. Smooth muscle fibers are not striated, hence their name. They contract at a slower rate compared to skeletal muscle, but can hold its contractions for longer amounts of time.

Muscle *

To build a cat you need muscles. Both bones and muscles are connected to tendons. The only action a muscle can do is contract. By contracting, the muscle pulls the bone that it is attached to by the tendons. To reverse this pulling, a different muscle needs to contract. This pair of muscles-called an antagonistic pair-pull the bone in opposite directions. A muscle is made up of many muscle fibers. These fibers are long, cylindrical cells, which are oriented parallel to each other. Each fiber is made of myofibrils, which are bunches of proteins such as like actin and myosin. Thin muscle filaments are made mostly of actin molecules, whereas thick muscle filaments are made mostly of myosin molecules. For the muscle to contract, thin filaments need to slide along thick filaments. To do this, myosin heads interact with the actin filaments by "walking along" them. They do this by binding and unbinding to the actin filaments. This process of contracting the muscle takes energy provided by ATP.

Nervous tissue *

Nervous tissue reacts to stimuli and sends information quickly. It is found in the brain, the spinal cord, and nerves throughout a cat's body. The units which make up nervous tissue are nerve cells, called neurons. The neurons conduct nerve impulses, and are specialized to do so. The neuron consists of three parts: the cell body, dendrites, and axons. The cell body contains the nucleus and organelles, and receives nerve impulses. Dendrites also receive nerve impulses. Axons can be bundled together to create nerves. Axons also transmit signals to other neurons or effector cells which can respond to stimulus. Neurons are not common compared to nearby cells. Some supporting cells surround and insulate the neurons, helping signals transmit faster, while others nourish the nerve cells and regulate the fluid around neurons. These neurons form nervous tissue.

Neuron structure / charge across membrane *

Neurons play an important role in the brains and nervous systems of cats. They transmit the tiny electrical signals which allow control of cognitive function, control of muscles, and functioning of senses. Individual neurons have central cell bodies, numerous, usually short extensions called dendrites, and a single long extension called an axon. Signals are sent to other neurons along the axon, and received by the dendrites or the cell body. An electrically insulating fatty substance called myelin is wrapped around the axon body in sections. These sections are called myelin sheaths. They serve to insulate the axon, allowing signals to travel much faster. The signals themselves are made possible due to the neuron's membrane potential. When a cell has a membrane potential, a charged substance (in this case sodium ions) is pumped (by a sodium-potassium pump) across a membrane, up its concentration gradient, producing a charge imbalance. Neurons build up this imbalance, then quickly open channels to allow the sodium to flow back into the cell when a signal is received. This rapid change in polarity is what carries the signal through the neuron. As mentioned above, neurons use sodium-potassium pumps to create a charge across their membranes; these specialized proteins pump three positively charged sodium ions out of the cell and two positively charged potassium ions into the cell every cycle. The result is a slightly higher positive charge outside the cell than inside.

Synapse *

In all animals, nerve cells must be able to transmit and receive nerve signals to/from other nerves. For this to happen nerve cells have an axon, a long extension used to transmit signals. On the end of an axon, there are branches of synaptic terminals. For the signals to jump synapses, relay points on each nerve, the synaptic terminal releases a neurotransmitter which stimulates the next cell. This process is repeated many times, very quickly so the animal can perform actions

Nerve impulse *

Before an impulse travels down the axon, the cell must set up an electrical current across the membrane by using the Sodium Potassium Na – K - ATPase pumps embedded in the cell's membrane. These proteins move three sodium ions out of the cell for every two potassium ions moved into the cell. This along with the higher concentration of chloride ions inside of the axon creates a negative charge inside the cell, compared to the outside. The voltage across the membrane is called resting potential. This sets up the axon for the next step.

When an impulse comes down the axon, sodium channels open, allowing sodium to rush into the axon. Then, the potassium channels open up and let the potassium ions flow with their concentration gradient, outside of the axon. This causes an action potential. The wave of depolarization moves down the rest of the axon, which can be up to a meter in length. At the end of the axon, the electrical signal causes the proteins attached to the cell membrane at the end to change shape, allowing synaptic vesicles to bond to those proteins. Synaptic vesicles carry neurotransmitters. Those neurotransmitters are secreted across the synapse, where they hit proteins on the main body of the next nerve cell. If enough neurotransmitters make it past the synapse and tell the next cell to keep propagating the signal, the signal will move on all the way to the brain. This is how the cat can respond to stimuli from the environment, such as feeling the floor it walks on or seeing the door right in front of it.

The cells must interact with their environment, which is still inside the cat. Things must be monitored, and certain items allowed into the cells, while certain items must leave. This is constantly regulated, and there are various mechanisms which allow the controlled movement of these items.

Passive transport *

Transport into or out of a cell without energy consumption is called passive transport, which occurs when a particle moves down its concentration gradient (from an area of higher concentration of the particle to an area of lower concentration) through random motion (diffusion). The main types of diffusion are simple and facilitated. Simple diffusion refers to hydrophobic (water-fearing) solutes which can pass through a membrane easily. Facilitated diffusion involves hydrophilic solutes which require a transport channel protein to allow the substance through a membrane, without the cell using any energy. Another type of diffusion which only relates to water is called osmosis. A cat will need passive transport to keep its cells isotonic, and will use diffusion and osmosis in every cell every day.

Osmosis *

Osmosis is the diffusion of water through a semipermeable membrane, meaning the membrane allows only certain things through. If there are different concentrations on either side of the membrane, osmosis occurs to make the concentrations the same on both sides. Water molecules surround the solutes on both sides. A lower solute concentration means a higher wat concentration. These water molecules tend to move across the membrane (through aquaporins), down their concentration gradient.

Osmosis occurs across the cell membrane, making sure the cat's cells do not become hypertonic or hypotonic relative to the surrounding fluid. If a cell becomes too hypertonic it will take in too much water, and explode (lyse). If a cell becomes too hypotonic, the cell will lose too much water, and will shrivel up. To prevent this, the cell goes through osmoregulation, keeping the cell isotonic. This means the concentrations of water and solute are equally concentrated on either side of the membranes.

Membrane transport proteins *

To build a cat from scratch, you will need membrane transport proteins, which play a big role in all animal cells. These membrane transport proteins act as a traffic cop in the cell and filter diffusion through membranes. Molecules such as O_2 and CO_2 get through the membrane without a problem but certain molecules and ions will require the assistance of these membrane protein "doors". The process of assisted transport by these proteins is called facilitated diffusion which is a type of passive transport because it consumes no energy. The proteins will help a certain substance that diffuses down its concentration gradient, thus requiring no energy. Some transport proteins will bind to their molecule and change shape to spill it out on the other side. Sugars, amino acids, ions, and water all use facilitated diffusion for crossing cell membranes. Water is a trickier substance to diffuse because of its polarity. For water to get through a membrane it must diffuse across a special protein called an aquaporin. Aquaporins have been found in bacteria, plants, and animals, and they will allow water molecules to pass through them.

Membrane transport proteins are essential to the cells of any living organism to get even the most basic things across the membrane. These proteins make for a smooth flow of molecules into the working cell.

Active transport *

A cell needs to use energy to move a solute, or something which that can be dissolved, against (up) its concentration gradient. The energy needed to move a molecule up its gradient is given by a chemical called ATP, or adenosine triphosphate. Active transport proteins, the proteins involved in this process, will only let certain ions or molecules into and out of cells. This helps the cells keep the concentration of ions and small molecules on the inside of a membrane higher than on the outside, in many cases. For example, on the inside of a cell there is a higher concentration of potassium ions (K+) than on the outside and there is a higher concentration of sodium (Na+) than on the inside. For sodium and potassium ions, there is a transport protein which that functions like a pump, moving potassium ions in and sodium ions out. Transport proteins work in the following process. First, a solute connects to the transport protein. Next, ATP provides the energy needed to change the shape of the protein to release the solute to the other side of the membrane. Finally, the protein returns to its original shape, allowing more solutes to attach and for the process to continue. It is important to have different concentrations of these ions for, among other things, nerve signal generation.

Endo/exocytosis *

Endocytosis and exocytosis are essential processes used to transport large molecules into and out of cells. Without these processes, the cells could not function properly and the cat would die. Endocytosis (*endo*, meaning inside) is the way some cells transport large items into the cytoplasm, which is the main fluid inside a cell. There are two types of endocytosis: phagocytosis (meaning cellular eating) which is how some cells "eat" particles by enclosing them in extensions of the cell membrane, called pseudopodia. The cell membrane is the outer perimeter of a cell that separates it from everything else. Once it has fully surrounded the particles it then transports them into the cell in a membrane-enclosed sac called a vacuole. (A cat's white blood cells of a type called macrophages performs this with bacteria, for example.) This vacuole then attaches to a lysosome which carries specific enzymes which flood into and digest the contents of the vacuole. The other type of endocytosis is called receptor mediated endocytosis. This is when the cell membrane forms a pit which is then coated with different receptor proteins, these proteins collect only specific molecules. Once this pit is filled with only the specific molecules, it closes off and releases these molecules into the cytoplasm.

Exocytosis (*exo,* meaning outside) is how some cells export bulky materials such as proteins or waste matter from the Golgi apparatus. The Golgi apparatus is where the cell changes, packages, and places proteins and other materials into little transport vesicles and then sends the vesicles to various points in the cell. The Golgi apparatus is the so called "post office" of the cell. So, once a transport vesicle is filled with the materials, it is transported toward the cell membrane. Once there, the vesicle fuses with the cell membrane and dumps its contents outside the cell. This is how some cells dispose of their waste and also how they export different useful molecules to be used somewhere else in the body. A nerve cell uses exocytosis to stimulate the next nerve cell in the line.

Cells comprise organs, which in turn make up organ systems. The cat itself must be a coordinated, integrated whole, possessing interconnecting systems. This is also true of the apple tree and the mouse.

(First, the cat and the mouse...)

Overview of food processing *

In a cat, food needs to be digested and processed so that the cat can receive the energy and nutrients it needs. When a cat ingests, or eats food, this is only the first step of four in breaking down what it eats into basic components that it needs and later uses. The cat needs to eat the right type of food to stay healthy. There are four main types of food molecules the cat requires: proteins, carbohydrates, nucleic acids, and lipids (fats). When a cat ingests food, it first chews and tears the food into smaller pieces with its teeth. In the second phase of digestion, food is chemically broken down by enzymes in the stomach lining and the small intestines. Animals' bodies cannot utilize foods until the molecules are small enough to pass through cell membranes. Through digestion, food is broken down from polymers, (chains), into monomers, molecules that bond together to form these polymers. This allows the "building blocks" of the foods (proteins, carbohydrates, nucleic acids and fats) to be split into small enough pieces to be absorbed into the cell membranes. The animal later uses these building blocks to make macromolecules that are needed in the body. The third step of food processing is absorption, in which cells that line the digestive tract absorb these building blocks into the circulatory system. The molecules then move through the body until they may be utilized to create new macromolecules, or may be broken for energy. The fourth step of the food processing system is elimination, when the animal gets rid of undigested or excess material. This leaves through the digestive tract.

Mouth/swallow/peristalsis *

An alimentary canal and accessory glands compose the digestive system of a cat. The oral cavity, or mouth, of the cat carries out mechanical digestion as it starts the entire process of digestion. As food enters the mouth, the cat's teeth divide it into small pieces. Its premolars and molars grind the food after it is being ripped apart by canine teeth and incisors. Small pieces of food enable the cat to swallow easily and expose largest surface areas of the food, maximizing the functions of the digestive enzymes in its mouth. At the same time, salivary glands transfer saliva into the oral cavity through ducts, stimulated by the presence of food. Substances in the saliva assist a smooth digestion. Bacteria entering the mouth is eliminated by the antibacterial agents in the saliva, while mucus, one of its other components, make food slippery and prepare it for swallowing. After saliva has functioned on the food, the cat's tongue shapes it into a spherical chunk, forming a bolus. The tongue then pushes the bolus into the pharynx. It connects both the cat's esophagus, a muscular tube leading to the stomach, and its trachea, a component of the respiratory system which functions as a windpipe. At this point, the process of swallowing begins. The swallowing reflex first takes place through the coordination of the epiglottis and esophagus. The epiglottis is a flat and elastic piece of cartilage, locating above the opening of the trachea. When no food is entering the cat's digestive system, it is contracted, pointing almost upright above the entrance of the trachea to allow the flow of air. When bolus enters the pharynx, the epiglottis relaxes to cover the opening of the trachea, preventing any food from slipping into the respiratory system. As food falls to the opening of the esophagus, the sphincter around it relaxes, allowing the food to go down. After the food has bypassed the sphincter, it contracts above the food and the epiglottis to its original position, reopening the entrance of trachea as it raises up. The process of peristalsis meanwhile pushes the food in the esophagus to the stomach: the ring-like smooth muscles of the esophagus involuntarily contract and relax in a wave-like motion. The muscles below the bolus first relax and open a pathway for the food. When the food slides down, the muscles above it contract and squeeze it down into the passageway below which is relaxed. Such contractions of the strong, elastic muscles of the cat's esophagus push the bolus down to the stomach in one direction.

Stomach *

Stomach is an organ which is part of the cat's digestive system. It has an elastic wall so it can stretch. In the stomach, chemical digestion happens. Cells in the stomach lining secrete acidic gastric juice. The stomach is protected from this acid by mucus. The gastric juice is acidic enough to inactivate most bacteria.

The gastric glands contain three types of cells which produce different components of gastric juice. The first are mucous-producing cells. They secrete the mucus which protects the cells which are lining the stomach. The second cell-type is the parietal cell. These secrete hydrogen and chloride ions which form hydrochloric acid. The last type is chief cell which secretes pepsinogen.

When pepsinogen and HCL meet in the lumen of the stomach, pepsinogen become pepsin. Active pepsin works to split polypeptides to smaller ones.

Mucus protects the stomach from pepsin and acids. The stomach lining constantly erodes, but new cells replace it completely every three days. Also, gastric juice is not constantly secreted. It is secreted when you smell or eat food. Nerve cells signal the brain and stimulates stomach. If food arrives in stomach, it releases gastrin which stimulates the secretion of gastric juice.

When food is present and stomach starts contracting, foods becomes a nutrient rich slurry, called chyme. This is then transferred to small intestine.

Small intestine *

The purpose of the small intestine is to digest and absorb food. The small intestine is lined with petite, raised tissues, which increases its surface area. This allows the small intestine to process nutrients much faster, because the tissues absorb nutrients. It processes around 90 percent of the food we eat. The small intestine is actually quite long, but its diameter is less than that of the large intestine. The small intestine is folded and wrapped in the abdomen. It is made up of three sections. The first of which, the duodenum, connects to the stomach. The duodenum carries out most of the chemical digestion. It then connects to the jejunum, which then connects to the ileum. The ileum is attached to the colon, or large intestine.

The pancreas and liver *

The main purpose of the pancreas is to make pancreatic juice which is a mixture of digestive enzymes (to digest food) and bicarbonate, which is an alkaline solution needed to neutralize the acidic chime coming from the stomach. The pancreatic juices are needed to break up and digest the chyme, a mixture of food and juices created in the stomach. Basically, without the pancreatic juices, the cat would not be able to digest any of the food it needs for energy. The pancreas also makes insulin and glucagon, which regulate glucose levels in the blood.

The liver produces bile, a complex mixture of acids, salts, phospholipids, cholesterol, water, and pigments. Bile acts as a detergent, and breaks fats to small droplets. This increases the fat's surface area, allowing lipases to more efficiently break the fats to components. This solution is stored in the gallbladder.

Large intestine and the evolutionary adaptations related to diet *

The large intestine follows the small intestine, and is wider in diameter. Its function is to process what remains after the cat's body has absorbed the nutrients. The cecum is a small bag connecting the end of the small intestine and the start of large intestine. It receives remaining waste from the small intestine, and lubricates it. The appendix is a small and thin, extra part, of cecum. Since the appendix contains white blood cells, it contributes to the functioning of the immune system.

The main portion of large intestine, which starts from the cecum and ends at the rectum, is called the colon. It continues the process of reabsorption of water started in small intestine. 90% of water in digestive juice is absorbed back to the body. The waste after the process of large intestine is called feces. They are more solid and most likely to be plant fibers. , which the cat cannot digest. The rectum is where feces are stored; and it is the final part of the colon. Two rectal sphincters are responsible for the operation of the anus. When the external sphincter, which is voluntary, relaxes, the rectum will dump out the feces.

In animals, evolutionary adaptations relating to diet are required for survival. Carnivores have big and expandable stomachs. Herbivores tend to have relatively long alimentary canals. A mouse would need longer digestion time to absorb more nutrients, since vegetables are harder to digest. Because animals do not have enzymes to digest the cellulose in plants, they need cellulose-digesting bacteria. A mouse has this type of bacteria in both its large intestine and its cecum.

Respiratory system *

The structure of cat's respiratory system is similar to an upside down tree. Water enters the tree through the bottom of the trunk, goes deeper and deeper into the branches, and finally reaches the leaves. Air in the cat's body generally follows the same process. There are two choices for air to enter cat's body, the nasal cavity or the mouth. After this, the air goes to the pharynx, the gateway for air and food. Next it is it's the larynx, or voice box. In this voice box, there are vocal cords, the musical "strings" which that form meows or yowls. Air then flows to the trachea, which is similar to the trunk of the tree. Just like the tree branches extending from the trunk, the trachea splits into two bronchi, each of which leads to a lung, where they will branch continually and repeatedly into thinner and thinner tubes called bronchioles. At the end of bronchioles are clusters of air sacs called alveoli, similar to leaves hanging on tree branches. Every cat has millions of alveoli in each lung, which creates an enormous surface area. On the inner surface of each alveolus is a thin layer of epithelial cells. The O2 dissolves in a film of moisture on these cells. A dense web of blood capillaries surrounds alveoli to exchange for gas.

Negative pressure/breathing control *

Breathing is the ventilation of the lungs by inhaling and exhaling. In cats, this occurs via negative pressure breathing. The key to this is creating a pressure gradient by expanding the lungs when inhaling, by contracting the rib muscles and diaphragm. These contractions cause the lungs to expand, making the air pressure in the lungs to drop below the pressure that of the atmosphere. Air going from a higher to lower pressure region is pulled through the nostrils into the lungs. The opposite of this happens for exhalation, reversing the gradient.

The diaphragm and rib muscles relax, causing the lungs to relax. The increased air pressure forces air back out of the body, allowing for the process to begin again.

Breathing is also involuntarily controlled for the most part, though you can hold your breath or hyperventilate voluntarily. The "control center" for breathing is a part of the brain called the medulla oblongata which that makes sure you intake enough air. pH changes in the cerebrospinal fluid bathing the brain informs the control center to increase breathing rate, bring the pH to a more normal level. Processes such as like these and others make sure breathing rate is kept at a rate appropriate to your body's metabolic needs. Now, it is it's time to look at how gases can be transported and exchanged.

Transport of gases in body *

Blood transports respiratory gases. One side of the heart handles oxygen-poor blood. The other side of the heart deals with blood rich in oxygen. The oxygen-poor blood comes back to the heart from capillaries in body tissues. The heart then pumps this blood to the lungs' alveolar capillaries in the lungs. Gases are exchanged between air in both the alveoli and the blood in the capillaries. The blood picks up O_2 and drops off CO_2. It then goes back to the heart and gets pumped out to body tissues, where it will drop off O_2 and pick up CO_2. The gas exchange between capillaries and their surrounding cells occurs by the diffusion of gas down gradients of pressure. A mixture of gases exerts pressure. Each type of gas has a partial pressure, which means it accounts for a part of the mixture's whole pressure.

Most animals transport O_2 bound to proteins called respiratory pigments. Cats use hemoglobin, a pigment containing iron which turns red when it bonds to O_2. Every red blood cell has about 250,000,000 hemoglobin molecules inside of it. These molecules are made up of four polypeptide chains. Attached to each polypeptide is a chemical group called a heme, the center of which is an iron atom. Since every iron atom binds to one O_2 molecule, a hemoglobin molecule can carry up to four O_2 molecules. Hemoglobins load up with O_2 in the lungs, and then transport it to body tissues. The amount of O_2 it unloads depends on the oxygen need of the cells. As blood flows through capillaries in the lungs, carbonic acid is formed. This turns into CO_2, which diffuses from the blood to the alveoli, and leaves the body through exhalation.

63.Circulation overview *

The circulatory system is an essential component of the cat's body. It allows for the cat to provide nourishment for itself as well as fight diseases, excrete cellular waste and maintain homeostasis. All types of bodily nutrients, such as amino acids and electrolytes, in addition to oxygen, carbon dioxide, hormones, nitrogenous wastes and blood cells are transported through the circulatory system.

In addition, circulation is the major proponent of homeostasis, or the stable equilibrium of the body.

Cats have a closed circulatory system, in which the blood always stays within the capillaries, veins, and arteries. Major components of the cat's circulatory system include the blood vessels, heart, and the network of veins, capillaries, and arteries which carry the blood (a fluid connective tissue). The heart is largely regarded as the "control center" of system of circulation.

Heart contraction and the SA node *

The cardiovascular system is composed by a circulatory system of capillaries, veins, arteries, and the most essential organ to build a cat, a four-chambered muscular heart made of cardiac muscle tissue. The heart is a transfer station of oxygen-rich and poor blood, where it receives oxygen-poor blood, pumps it to the lungs, receive oxygen-rich blood from the lungs and pump it back to the body. The four chambers of a heart consists of a right ventricle, a left ventricle, which are below the left and right atria. The ventricles have much thicker walls than the atria because they need to pump blood to the lung and the body, unlike the atria which that only squeeze blood into the ventricles. There are valves at the exit of every chamber to keep the blood flowing in one direction in the heart, and the valves which that connects the ventricles and the atria are called atrioventricular valves. The rhythm which a heart beats is called the cardiac cycle. It contains two phases: diastole, when the heart relaxes and blood flows into the heart; systole, when the heart contracts at first by the atria, then by the ventricles and pump the blood out to the body. The amount of blood per pump is called cardiac output and the beats per minute is called heart rate. The heart doesn't pump according to the signals from the nervous system. The cat's heart utilizes a sinoatrial node. It set up the pace which the heart beats by sending signals to the heart similar to the manner in which nerve cells function. The signals are first sent to the atria to coordinate their contractions. Then the signals go to the atrioventricular node. After a delay to empty the atria, the signal reaches the ventricles finally and triggers the contraction. The signals are strong enough to be detected on the skin of the cat and determine whether there are health problems in the heart.

Blood vessel structure *

The circulatory system consists of the the heart, blood, and of course, the blood vessels. In order to have a living cat, it must have numerous sets of tubes to carry blood to the heart, from the heart, and throughout the body. This can be accomplished through the functions of blood vessels. Varying in structure and function, each category of vessels are responsible for fulfilling their unique task. To list a few, arteries, arterioles, veins, and venules are all different categories of blood vessels, each differentiating in size, structure, and location relative to the heart.

Before one begins building the blood vessels, it is important to understand the structure of a cat's heart. Like all mammals, the heart of a cat has four chambers. Two of the four chambers are the atria, which are not blood vessels themselves. Rather, they are small and hollow structures. Ventricles make up the second half of the heart, pumping out the blood given to them from the atria. Large arteries then carry blood out from the each ventricle into further vessels. Now that the structural overview of the heart is understood, the blood vessel building process can begin.

Smooth epithelial walls, and two other layers of walls, are critical to the building process of the vessels. The first step is to craft the outermost layer of tissue, or the tunica externa. Made up of elastic fibers, one may begin binding the fibers into a tubular shape, making sure it has the largest circumference out of the following layers. After which the task is complete, one can begin molding the tunica media, making the circumference smaller than the tunica externa. Lastly, begin the inner layer of smooth muscle, or the tunica intima. It is crucial that this layer is smooth. If the surface is rough, the risk for plaque buildup in the vessels is highly likely, leading to atherosclerosis. The only variable in this process is each blood vessel's location and size, which is explained in the next steps.

As mentioned before, large arteries carry the blood away from the ventricles. Construct the larger arteries to extend out of the heart to further lead to a smaller blood vessel. Before making the smaller blood vessels, one also needs to make veins. Repeat the steps in the making of the arteries, except add a valve inside of the innermost wall (epithelium). These valves acts as flaps so blood will travel to the heart and only the heart. The walls don't need to be thick, or elastic, as they only deal with low pressures.

Next, the smaller blood vessels branched off from the arteries. To build these, make sure they branch off the sides of the large arteries. They should bring about the capillary beds. Manufacture the capillaries themselves before linking them all together to create the bed. Capillaries are much smaller than the arteries and venules. Lastly, they should merge with venules. Capillaries are one cell this, which allows things to move from blood to tissue, and from tissue to blood. They serve a different purpose, though. Instead, they are low in oxygen concentration and connect to the veins. By now, most types of the blood vessels should have the three layers and be linked together, from the atria, through the capillary beds, and back to the ventricles.

Capillary Function *

Capillaries are the smallest blood vessels, as they are
only one cell thick! The capillary wall is made of
epithelial cells. Because the walls are so thin, they are
permeable to gases, nutrients, and waste. Small
molecules such as O_2 can directly diffuse through the
walls, but larger molecules may need to be carried by
vesicles through endocytosis and exocytosis.
Additionally, there are pores and clefts in between the
cells. Water and other small materials can move
through these openings. White blood cells can also
get through these cracks because they are
amorphous and can thus change into a thin shape.
However, red blood cells and dissolved proteins are
too large and rigid to exit the capillary. In fact, a
capillary is barely large enough for red blood cells to
move through it one by one! Most of the exchange of
materials between the capillary and interstitial fluid is
due to the flow of water and dissolved solutes
through the openings. Blood pressure pushes outward
into the interstitial fluid, while osmotic pressure pulls
back into the capillary. The cat needs these capillaries
to transfer cells and molecules in and out of the
circulatory system

Innate Immunity *

The immune system protects against pathogens
(disease producers). Innate immunity is a form of
defense that is always active. Skin is one such
defense, preventing pathogens from getting to the

bloodstream. Internal areas that must be exposed to the outer air have mucus-producing membranes to trap pathogens, and especially vulnerable areas have extra protection, such as the cilia that line the throat and force pathogens trapped in the mucus out. Microbes that get past those defenses are attacked by innate immune cells, which are called white blood cells, or leukocytes. The two most common of these are neutrophils, which locate sites of infection, and macrophages, which devour any pathogens they encounter.

Cells that become infected have defenses to keep the infection from spreading. When a cell is infected, it starts producing interferons, proteins which inhibit the growth and spread of pathogens. The cell itself dies, but the proteins it produces protect other cells from being infected.

Adaptive immunity *

A major component of the immune defensive system is called adaptive immunity, or acquired immunity. Our cat's adaptive immunity will be dependent on its individual exposure because adaptive immunity varies based on previous encounters with antigens. Antigens are molecules attached to the outside of pathogens which cause an adaptive immune response, and the immune system reacts to antigens by attacking them or releasing antibodies which correspond to the certain antigen. Antibodies are proteins released into the blood which fasten to antigens to counteract their impact. Adaptive immunity can respond more quickly and effectively to antigens to which it has been previously exposed, so if the cat has previously been exposed to an antigen, its body will react more efficiently the next time it encounters said antigen.

Adaptive immunity can be obtained in two ways: 1. Infection, and 2. Immunization. Infection is encountering antigens naturally, and immunization is where a vaccine is inserted into the immune system. The vaccine will contain an inactive strain or an inactive piece of a virus, but the body still manufactures antibodies which react to this type of virus. Obtaining adaptive immunity from infection or immunization is active immunity. If the antibodies are already manufactured elsewhere, but passed on to the cat, this is considered passive immunity. An example is when the queen (mother cat) feeds milk with antibodies she has produced and her kitten.

When the cat gets infected, the lymphatic system helps its body react by removing harmful agents and returning fluid to the blood. The fluid diffuses into capillaries, which then drain the fluid into lymphatic vessels. The lymphatic vessels transport fluid around the body, squeezed by the movement of muscles. These lymphatic vessels pass into lymph nodes, which hold the toxins collected from around the cats body. Lymphocytes and macrophages flood into the lymph nodes to fight these toxins. Stem cells from bone marrow develop into lymphocytes. Lymphocytes can develop into either B cells or T cells: B cells are lymphocytes which develop in the bone marrow and T cells develop first in the bone marrow but then move above the heart in the thymus. Both B and T cells have antigen receptors which can attach to certain antigens. Once the B and T cells develop their antigen receptors, they can move to lymph nodes and lymphatic vessels through the bloodstream. There are two responses which B and T cells have to antigens: the humoral immune response and the cell-mediated immune response. The B cells use the humoral immune response to secrete antibodies to defend against free-floating antigens in body fluids. The T cells use the cell-mediated immune response to destroy cells which have been already infected by antigens. Our cat's adaptive immune system will help it fight infections and thus increase its chances of survival.

Clonal selection *

Out of the pool of diverse antibodies, only a tiny percentage will ever be activated to fight the invading antigens. However, as one antigen triggers the B cell with corresponding antigen receptors, this specific kind of B cell will start to form identical clones of itself. This immune response is known as clonal selection. The proliferated B cells will divide into two kinds of cells: memory cells and effector cells. Memory cells are what gives a cat lifetime immunity against diseases. Effector cells, also known as plasma cells, fight against existing antigens immediately. While effector cells fight against antigens immediately by releasing antibodies into lymph and blood, memory cells remain in the lymph nodes. These memory cells in the lymph nodes will wait to be activated when the corresponding antigen invades the body again. This immune response not only protects the cat against invading antigens, but also prevents future infection caused by the same pathogen.

1' and 2' response to infection *

The primary response to an infection refers to the first time the body encounters the antigen and the cat's body activates lymphocytes (white blood cells). During the primary response, it produces B cells which will split into two different types, memory cells and effector cells. The memory cell will recognize the antigen and slowly produce specific cells which secrete antibodies into the blood and overpower the infection. This can take up to 2-3 weeks to take effect. The secondary immune response is much faster because the memory cells have already learned this specific antigen. Now the secondary response can produce the effector cells from its memory cells and destroy the antigen within a few days. So now, every time this specific antigen enters the body the effector cells can be produced immediately.

Antibodies *

When building your cat, make sure the cat is not affected by pathogens (a pathogen is basically anything that can cause disease). Antibodies will help keep the cat safe and protected from these pesky microorganisms by identifying and marking these invaders for extermination. Antibodies are basically formed of 4 chains of polypeptide. The function of an antibody is dependent on the stem of the antibody's distinct three-dimensional shape; this means that if the stem of the antibody has a different shape, it will be bound to a different pathogen. Antibodies are usually secreted by Effector B cells. These cells are incredibly powerful and efficient when battling existing infections in your cat's body.

You will need to make the antibodies for your cat in a "Y" shape to have areas to attach to pathogens and areas to destroy them. On the "Y", at the end of each arm, you must form an antigen-binding site. This area is responsible for the antibody's ability to recognize and bind to different pathogens. The stem of the "Y" helps moderate getting rid of the antigen once it is bound. In all antibody mechanisms, there are 2 important parts. One is a recognition and attachment phase while the other is a destruction phase. The recognition and attach phase is specific, as mentioned above. The shape of the antibody matches its function. The destruction phase is nonspecific however and is not determined by the shape or type of antigen.

T-cells *

For a cat's body to effectively combat pathogens, it must utilize its adaptive immune response. This destroys antigens in its blood, lymph or intestinal fluid. However, many invaders have already entered the body and reproduce in these fluids. Cytotoxic T cells produce a cell-mediated immune response which attacks these invaders that have already entered the body. But before the cytotoxic T cell can attack an invader, it must be stimulated by signaling molecules that are secreted by helper T cells. Helper T cells trigger humoral and cell-mediated immune responses. However, the helper T cells themselves do not carry out these responses. In order for a helper T cell to activate adaptive immune responses, there are two requirements that must be met. First, there must be an invader that can bind specifically to the antigen receptor of the T cell. Second, this antigen must be available on the surface of an antigen-presenting cell (e.g. Macrophages and B cells). If we consider a macrophage, there are three basic steps that it will go through in order to activate the helper T cell.

1. It ingests a foreign molecule and breaks it down into foreign antigens
2. The molecules of a special self-protein belonging to the macrophage bind to the antigen molecules, nonself-molecules, and displays them on the cell's surface
3. The helper T cells recognize and bind to the structure that has a foreign antigen and a self-protein, called a self-nonself complex, which was displayed on the cell's surface

The helper T cell's ability to specifically recognize a unique "self/-nonself" complex relies on the embedded receptors in a T cell's membrane. Each receptor has two binding sites: one for the antigen and one for the "self" protein. These binding sites enable a T-cell receptor to recognize the overall shape of the a "self/non-self' complex on an antigen-presenting cell. This binding of a T-cell receptor to a "self/non-self" complex prompts the activation of the helper T cell. The secretion of stimulatory proteins by these T cells have three major effects

1. They stimulate the development of memory cells and additional effector helper T cells
2. They help activate B cells, which activates the humoral immune response
3. They stimulate cytotoxic T cells

The other type of T cell is the cytotoxic T cell, which actually kills the infected cells. A T cell recognizes an infected cell by analyzing the foreign antigens on the surface of the cell. There are five basic steps in the process of how a T cell destroys an infected cell.

1. A cytotoxic T cell binds to the infected cell
2. This stimulates the T cell to create toxic proteins that act on the infected cell
3. Perforin (a toxic protein) is discharged from the T cell and it makes holes in the infected cell's plasma membrane
4. T cell enzymes initiate the death of the cell via apoptosis, which causes the cell to lyse

5. The cell is destroyed, and the cytotoxic T cell moves on to take care of other infected cells.

Thermoregulation *

Thermoregulation is the maintenance of a steady internal body temperature under various external conditions. Cats rely on thermoregulation to keep their body temperature stable, allowing them to perform enzyme-mediated processes effectively, even when the external conditions vary. Cats and other types of mammals are endothermic, which means they are warmed from the heat generated in metabolic reactions. Endotherms can modify their metabolic heat production using both shivering and non-shivering thermogenesis. An important example of shivering thermogenesis is muscle contraction. During muscle contractions heat is released from the chemical reactions powering the muscle contractions and the friction produced between muscles fibers. Non-shivering thermogenesis is primarily located within brown fat or brown adipose tissue. Brown adipose tissue Brown fat contains many mitochondria which have specialized proteins that directly release energy from molecules as heat instead of ATP.

Another adaptation cats utilized to alter their heat production is vasoconstriction and vasodilation. Vasoconstriction causes the surface vessels to become constricted , creating less blood flow between the cat's core to its surface and reducing heat loss. During vasodilation, the surface vessel expand creating more blood flow and an increased rate of heat loss. Also, Cats use countercurrent heat exchangers to allow heat to be transferred from warmer blood to colder blood in the circulatory system. In addition, most cats have a layer of fur, which reduces radiation from the cat to its environment.

Osmoregulation *

Osmoregulation controls the concentration of solutes in water. If a solution surrounding a cell becomes hypotonic, the water diffuses into the cell causing the cell to burst. However, if a solution becomes hypertonic, the water will diffuse out of the cell causing it to shrivel up. The best level is if it remains isotonic, where the water concentration remains constant in and out of the cell. This means in our cat we need solutions to be isotonic. Cats are osmoregulators, meaning they must actively regulate their water levels so they are able to maintain homeostasis. Animals are able to gain water through drinking, eating, and cellular respiration.

Much osmoregulation occurs in our cat's urinary system. The main organs of the urinary system are the kidneys. They concentrate the urea, while reusing the water and helpful nutrients by returning them to the blood after having filtered them out. This is the filtration process. The product of filtration is known as filtrate.

After the filtration into the kidneys, roughly 65% of the water is reabsorbed and follows the solutes through the nephric tubule. More water reabsorption is allowed throughout the rest of the filtrate's journey through the kidney because of the solute gradient that is maintained. During the travels through the tubules osmoregulation allows water to diffuse into or out of a tubule. The filtrate travels from the proximal tubule to the loop of Henle, distal tubule, ending in the collecting duct. This is where it connects to the ureters, so the cat can dispose of the urine.

Hormones play a key part in regulating the urinary system. Keeping the homeostasis between the water and solute is controlled by the release of hormone (ADH) from the cat's pituitary.

ADH *

ADH (antidiuretic hormone) from the pituitary acts on the kidney's collecting ducts to maintain the water and solute levels. When the cat is dehydrated, the solute concentration of its body will get very high, so the brain increases the level of antidiuretic hormone (ADH) in its blood. The cells in kidney's collecting ducts are then able to reabsorb water from the filtrate. The reabsorption of water increases the amount of water in the cat's blood, thus increasing the blood pressure.

However, if the at drinks too much water, its blood pressure increases. Consequently, the ADH will drop, which will cause the collecting duct cells to reabsorb less water and thus result in watery urine (diuresis). The cat's urine will then be very clear. Increased urine is diuresis, and ADH is called antidiuretic hormone because it acts against this state.

Chemical and Electrical Signals *

The cells of an animal's body must communicate to maintain homeostasis, as well as to carry out coordinated functions with each other. They do so using chemical and electrical signals running through the body's two main communication systems: the nervous system and the endocrine system.

The endocrine system is a group of interacting glands and tissues throughout the body of animal. The glands involved produce and secrete chemicals in order to coordinate body functions. These chemical signals are called hormones. The endocrine system is well suited to organize gradual changes which affect the whole body. Hormones also regulate long-term processes throughout the body. The organs which make and release hormones are called endocrine glands. The hormones travel throughout the body via blood vessels. Not all cells receive these hormone signals. Certain target cells which have receptors (target cells) for a specific hormone will respond to the signal.

The other communication system is the nervous system. The signals of the nervous system are mostly electrical, and are transmitted along nerve cells, (neurons). These signals travel along defined nerve pathways, rather than through the bloodstream. However, similar to hormones, these signals are also only traveling to specific target cells. There is always a direct connection between the neuron firing the signal and the target cell meant to receive it. The electrical signals of neurons, unlike hormones, last less than a second, and their effects are quick and not long lasting.

Hormone signaling mechanisms *

There are three stages the cat needs to go through to complete hormone signaling: reception, transduction and response. For reception to occur, a hormone needs to bind to a certain receptor protein inside or outside a target cell. Transduction takes place inside the target cell and converts signals from the hormone to receptor proteins. The final stage is response, which is when the cell's actions change. Hormone groups can be classified into water-soluble hormones and lipid-soluble hormones. These each go through the three stages of hormone signaling but do so differently.

A water-soluble hormone cannot travel through the cell's phospholipid bilayer, so their receptor proteins are bound to the outer (plasma) membrane of the cell. For the first step, the cat needs the hormone to attach to the receptor protein, and stimulate it. The receptor protein will signal a transduction pathway made up of relay molecules, which converts the signal into a form that the cell will use. Lastly, the relay molecule stimulates a protein which will initiate the response of the cell. These responses can either be executed in the cytoplasm for stimulating an enzyme or in the nucleus to regulate gene expression.

Lipid-soluble hormones, however, pass through the phospholipid bilayer by diffusion, so the receptor proteins are located within the cell. An example of a lipid-soluble hormone is a steroid hormone. Let's follow the steroid hormone through the steps. In the first step, the steroid hormone attaches to a receptor protein that can be located in one of two places, the cytoplasm or the nucleus. No relay molecules, therefore, are required. A hormone-receptor complex completes the transduction of hormone signals. Then, the complex binds to a specific site on the DNA in the nucleus and this activates gene regulation by switching genes on or off.

Signal transduction *

Signal transduction enables cells to communicate messages to coordinate gene expression, and is very important for coordinating cellular activities. Usually, you will need another cell which sends a signaling molecule. The molecule then will bind to a receptor protein located on the plasma membrane of the target cell. A transduction pathway is initiated in the target cell. The many relay proteins inside the target cell, will activate one another in sequence, forming the transduction pathway. The final relay molecule activates a transcription factor, allowing a specific gene to undergo transcription.

Vertebrate endocrine system *

The vertebrate endocrine system consists of more than a dozen glands, each of which has essential functions. Some of these glands work specifically in this system, such as the thyroid, whereas others, such as the pancreas, have multiple functions, some endocrine, some not. The main job of the endocrine glands is to produce hormones. There are many different ways a gland is stimulated to produce a hormone. One way involves a change in ion or nutrient levels. The gland thus stimulated will produce a hormone to stabilize these levels. Another way is direct stimulation by the nervous system. This is how the cat's adrenal glands are stimulated and produce adrenaline. The last main way these glands may be stimulated is by other glands releasing hormones. For example, the anterior pituitary is stimulated by the hypothalamus releasing hormones. Next, we will talk about the effects hormones have in your cat's body.

The hormones, depending on the ones released, have a wide range of effects in your cat's body. Hormones can help maintain metabolic regulation, ion/nutrient regulation, stabilize growth, reproduction, development, and can be released in response to certain emotions.

For your cat to reproduce, it needs the relationship between the hypothalamus, the pituitary gland and the ovaries. The hypothalamus releases the follicle stimulating hormone which stimulates the follicle growth. The follicle contains the egg. The ovaries then release estrogen or progestin and signal the hypothalamus to release the luteinizing hormone to signal the ovaries to begin ovulation, which is release of the egg. Through this cycle between major endocrine glands, your cat will be able to reproduce. Hypothalamus *

The hypothalamus is a fundamental part of a cat's endocrine system. It receives and responds to the information sent by the nerves. The hypothalamus acts as the headquarters of the endocrine system. The pituitary gland, attached to the hypothalamus, sends chemical instructions to other glands and organs. It is made up of a posterior lobe consisting of nervous tissue, and an anterior lobe consisting of endocrine cells. The hypothalamus is connected to the posterior pituitary by a set of neurosecretory cells. These neurosecretory cells build some of the hormones. The hormones are transported throughout the cat's body by blood vessels. The anterior pituitary also makes and releases hormones. It is connected to the hypothalamus through two types of hormone, the releasing hormone and the inhibiting hormone. The releasing hormones cause the release of other hormones, whereas the inhibiting hormone stops the secretion of other hormones. Thyroid-stimulating hormone (TSH), for example, balances the level of thyroid hormones released into the cat's blood. Prolactin, a hormone of the anterior pituitary, allows the cat to produce milk by triggering the mammary glands. Growth hormone stimulates the synthesis of proteins and promotes growth. (In humans, giantism is a result of a very high level of growth hormone whereas dwarfism is caused by an insufficient amount of growth hormone. The hypothalamus releases TSH-RH, or Thyroid stimulating hormone-releasing hormone. The cat's metabolic rate is increased by thyroid hormone. The increase of metabolic rate warms the body. During cold temperatures, more TRH is secreted. This, then, is part of thermoregulation.

Thyroid *

The thyroid gland is wrapped around the trachea and it produces and releases two types of thyroid hormones: triiodothyronine (T3) and thyroxine (T4). These two hormones control the rate of metabolism. The thyroid hormones are modified amino acids, bonded to three or four iodine atoms. The thyroid hormones are controlled by end-product inhibition in which the final product acts as an inhibitor for one of the steps in a series of reactions. If the thyroid gland can't synthesize thyroid hormones because of the lack of iodine in the cat's diet, goiter, a swelling of the thyroid gland, may result.

Sex hormones *

Sex hormones, produced in the gonads, help in the growth and development of the body. These gonads produce gametes: sperm for male and ova for female. The female reproductive organs are the ovaries. Male reproductive organs are the testes. The main reproductive hormones are estrogens, progestins and androgen. The estrogen, primarily found in the female cat, functions to regulate and maintain their reproductive system. Progestin, or progesterone, prepares and maintains the uterus and embryo in in the female (queen). Androgen, on the other hand, is responsible for the characteristics of the tomcat.

Pancreatic hormones *

The pancreas is a gland that has two major functions, it secretes digestive enzymes into the small intestine, and it secretes insulin and glucagon, hormones, into the blood. This regulates how much glucose is found in the cat's bloodstream. In the pancreas, there are clusters of endocrine cells, known as pancreatic islets. Each islet contains beta cells which produce insulin, and alpha cells, which produce glucagon. Insulin and glucagon are known as antagonistic hormones, as the effects of one works in opposition to the other. These two hormones balance a "set point" of glucose. There are two negative feedback systems which also help manage the amount of glucose found in the cat's bloodstream. One of them releases insulin into her bloodstream, lowering the amount of glucose, while the other one releases glucagon into the bloodstream which raises the level of glucose. When insulin is present in the cat's bloodstream, liver and muscle cells take up glucose. And convert it to glycogen. The amount of glucose in the cat's bloodstream will decrease as the cat uses energy. In response to this decrease, glucagon is released to the blood, and the liver and muscle cells break down the stored glycogen, releasing glucose into the blood. Now, the amount of glucose present in the bloodstream will increase.

Adrenal gland *

What happens when a cat gets scared and runs away? Its heart rate and stress level increase. This is caused by the adrenal glands, an endocrine system gland which sits above the kidney and produces hormones which respond to stress. This is very important for a cat when in dire situations. The adrenal gland is composed of two parts, the adrenal medulla and the adrenal cortex, so be sure to have two parts prior to constructing the cat.

The adrenal medulla, the center portion, is responsible for the "fight-or-flight" response, rapid and short term. Make sure it secretes two different hormones, epinephrine, (adrenaline), and norepinephrine, (noradrenaline), which activate near a nerve cell and are secreted into the blood. Both these hormones will work to release glucose into the blood and help prepare the body for raised heart rate, blood pressure and metabolic rate. Also, make sure the adrenal medulla can either constrict or change blood flow.

Next, the adrenal cortex, is the outer portion of the gland and has a slower, long-term response to stress. Make sure it responds to the chemical signals in the blood by producing corticosteroids, a steroid hormone. There are two main types of corticosteroids: mineralocorticoids and glucocorticoids, so make sure these are being produced. Mineralocorticoids work mostly with water and salt balance. They should help the kidneys to absorb sodium ions and water, thus increasing blood pressure. Next, the glucocorticoids should mobilize cell energy reserves. By breaking proteins and fats.

Placenta *

The duration of pregnancy for our cat is 64-67 days. This process begins with fertilization. Fertilization usually occurs in the oviduct. The oviduct is in the female cat's abdominal cavity, between the ovary and the uterus. Within 20-28 hours of fertilization, the zygote (fertilized egg) divides into a two-celled embryo. About a week after fertilization, the embryo reaches the uterus and cleavage has, to this point, produced 100 cells. The embryo is now a blastocyst. The blastocyst has an outer layer of cells called the trophoblast. Eventually, the trophoblast cells form part of the placenta. As the embryo develops, so does the placenta. Eventually, the placenta develops and it provides our soon-to-be kitten with oxygen, nutrients and waste removal.

Reproductive system *

The reproductive system of the house cat has similarities to the human, but they are also very different. Female and male cats and humans alike have gonads (organs which that produce gametes which can unite with gametes produced by the opposite sex to form a fertilized egg). The female possesses ovaries (which contain follicles that hold eggs, or female gametes), while the male has testes, which produce sperm cells. In cats and humans alike the female reproductive system contains oviducts, which are the tubes an egg travels through to get to the uterus, as while as a cervix which separates the uterus from the vaginal opening. However, the reproductive cycle of cats is different than that of humans. Female cats have an esterous cycle, which allows them to come into heat twice a year, as opposed to the human female's menstrual cycle, which is continuous through a 28 day calendar.

Sperm formation *

Inside the testis of a male cat there are twisted seminiferous tubules, where sperm formation occurs. Near the outer wall of the seminiferous tubules, diploid cells start the process of becoming sperm cells. Diploid cells contain exactly two sets of chromosomes which is important for when the cell begins to split. These diploid cells go through mitosis, or cell division, several million times until the cell begins the first stage of another kind of cell division, called meiosis. During meiosis I, the diploid cell divides into two haploid cells, each with one set of chromosomes. Then each of the haploid cells splits into two once again, thus creating four haploid cells in total. These cells then mature to become sperm cells through cellular differentiation, a process which turns them into specialized reproductive cells, or gametes. Once a sperm is fully formed it is moved toward the center of the seminiferous tube. The entire process may take weeks.

Oogenesis *

Cats are created by the combination of a sperm cell and the ovum, forming an embryo and eventually developing into a mature being. Oogenesis is the name for the formation of a mature ovum.

Before birth, there is a diploid cell which contains all 38 chromosomes. After a series of preparatory steps for meiosis I, the diploid cell become a "primary oocyte" and is ready for the hormone signal to start meiosis I.

Meiosis I splits the cell into two parts, each containing 19 chromosomes. However, the division is not equal. One of the two cells contains most of the cytoplasm, whereas the other has almost no cytoplasm. The latter is called the first polar body, and it degenerated after the division. The other cell develops into the "secondary oocyte", waiting the entry of sperm before beginning meiosis II.

Eventually, the sperm penetrates the secondary oocyte, which triggers it to divide into a the secondary polar body and the final product - the mature egg.

A follicle holds the ovum inside the ovary. During the primary oocyte period, the follicle grows continuously. When it becomes fully mature, the oocyte inside also completes meiosis I and becomes the secondary oocyte. It ruptures and sends the ovulated secondary oocyte outside the ovary, and it, itself, develops into a corpus luteum, or yellow body. The yellow body then either degenerates or, if fertilization occurs, releases hormones to maintain pregnancy.

Fertilization of Egg Cells *

For cats to thrive in the future, they need to be able to reproduce. All organisms reproduce to grow the population of their species. As in most species, males contain sperm cells which are used to fertilize the egg of the female. Of the millions of sperm which are ejaculated, only one will successfully fertilize an egg.

The shape of the sperm is very elaborate giving it the traits to complete this action. The sperm cells contain a long flagellum useful for swimming through the liquids in the vagina, uterus, and oviduct of the female to the egg. It has mitochondria in its structure to utilize fuel (fructose), and a head containing a nucleus and an acrosome at the tip of the head which will open up to the egg when fertilizing it. The steps in which the sperm fertilizes the egg are as follows. The sperm will first come into contact with the jelly coat of the egg which will trigger the release of an hydrolytic enzyme cloud from the acrosome. Tis forms a cavity in the jelly coat that allows the sperm to penetrate further into the egg. The sperm head will finally find its way to the vitelline layer. Here, species-specific protein molecules on the sperm's surface will attach to the receptor proteins on the outer egg, which makes sure no other sperm can fertilize the egg. These steps then form a diploid cell, a zygote, possessing chromosomes from both the sperm and egg. This will further develop through many stages to become an embryo that will grow into a complete organism.

Zygote/Blastula/Gastrula/Organogenesis *

A zygote is created when a sperm cell of a male cat and an egg of a female cat fuse. The zygote undergoes a series of rapid cell divisions called cleavage. Once the zygote divides for the first time it is called an embryo. As cleavage continues and the embryo divides into more and more smaller cells, and a small fluid filled cavity forms inside the embryo. This is called a blastocoel. The embryo becomes a hollow ball of many small cells and is called the blastula. Once cleavage is over, the rate at which the cells of the embryo divide slows dramatically. This is when the embryo undergoes gastrulation, which is the second major phase of embryonic development. During gastrulation, specific cells in the embryo move to new locations within the embryo which will allow for later development of all the different organs and tissues. As gastrulation continues, the embryo forms a three-layer stage called a gastrula. The three layers are all embryonic tissues, called the ectoderm, the endoderm, and the mesoderm. After gastrulation is completed, organ formation occurs. This is when the gastrula creates the different embryonic organs from the three embryonic tissue layers. These, then, are the first stages of growth in the actual building of a cat.

Cell signals direct the embryonic development *

Cellular differentiation is caused through selective gene expression. This process is most dynamic following the embryo's development from a zygote, which is a product of fertilization events. During the development of the embryo, the zygote communicates with adjacent cells. Signals trigger the expression of genes. Then, the zygote transforms into an embryo with repeated mitotic divisions. Cell signaling takes place among the cells of the embryo, causing the development of the embryo. In the next stage, the homeotic genes control the subdivision of the embryo's body into segments. Homeotic genes are a specific set of genes which determines and control which body part will develop from certain cell clusters. Mutation of the homeotic genes cause abnormal animals. One example of mutation resulted in the Scottish Fold, a breed of cat with folded ears.

Nervous system overview *

The cat's nervous system is made up of the brain, sense organs, spinal cord, and nerves. It coordinates activities of the body by sensing for stimuli (something which creates a reaction), using information, and responding accordingly allowing fast coordination of bodily functions. The nervous system is comprised of two divisions, the peripheral nervous system (PNS) and the central nervous system (CNS). The peripheral is made up of the neurons which transport the signals in and out of the central nervous system, which includes the is the brain and spinal cord. To communicate, the nervous system uses nerve cells, or neurons, which send information with chemical and electrical signals. The nervous system has three main jobs. First, there is *sensory input*, where the signals from sensory receptors are conducted. Then, *integration*, or the interpretation and analysis of the signals. In this phase, the system formulates how to respond to each signal. Finally, *motor output*. This is the movement of signals through the PNS from the centers of integration.

Vertebrate nervous system *

The peripheral nervous system (PNS) is one portion of the nervous system. This system consists of both sensory and motor nerves. Sensory nerves receive information from receptors and motor nerves send information from the central nervous system (CNS) to organs that respond to nerve impulses, or effector organs. The motor system transports signals from the CNS to skeletal muscles. Another component of this system is the autonomic nervous system, which regulates the internal muscles. This system is usually involuntary and the cat does it naturally. Within the autonomic nervous system, there are three divisions. The first division is the parasympathetic division, which prepares the body for activities that conserve or gain energy. The parasympathetic division is active when the cat is digesting something or resting. The second division is the sympathetic division, which prepares the body for intense activity that consumes energy, such as the cat chasing after a mouse, or fleeing a dog. The third division is the enteric division, which controls the gastrointestinal activity. This division can work on its own unlike the parasympathetic and sympathetic divisions, who rely on each other to function. Both parts, the motor system and autonomic nervous system, cooperate to maintain homeostasis within the body.

Brain *

The brain is a complex organ. This organ is made up of billions of neurons and is more sophisticated than a computer. It is made up of three major parts, the forebrain, midbrain, and hindbrain. Both the forebrain and hindbrain have more specific parts to it. The forebrain is the wrinkly and the biggest section of them all. Sophisticated actions take place on the surface of the forebrain, and so with more surface area (wrinkles), it can be able to perform more actions. It is responsible for higher order thinking. This part of the brain also contains the hypothalamus for example, which contains our "biological clock." It tells us when to wake up and when to sleep via the pineal gland, which releases melatonin when it is dark. The midbrain receives sensory data and sends it to the forebrain, and is also involved in reflexes. The hindbrain, at the bottom of the brain, controls the basic functions a cat needs such as breathing (pons and medulla oblongata), digestion (medulla oblongata), and swallowing (medulla oblongata). The hindbrain can be referred also as the "reptilian brain" because it controls such basic actions. This section is directly connected to the spinal cord.

Reticular formation *

The Reticular formation is a structure of the brain which involved with keeping the body alert, coincidentally the reticular formation is also involved in the process of sleep. It is a loosely connected formation which that extends through the brainstem. The reticular formation receives information and acts as a filter. It is in charge of the kind of information that gets to the cerebral cortex. As information sent to the cerebral cortex declines the cat's body becomes less alert. If a lot of information is going the the cortex the body will become more alert. Sleep, although important, is not a rest for the brain, the brain is still being stimulated during sleep. Electroactivity and brainwave activity continue working, especially during what we know as Rapid Eye Movement (REM) sleep, where brain waves are irregular and similar to when we are awake. This "awake-like" state of the brain is why we have the majority of our dreams during REM sleep.

A cat would obviously need such systems to ensure it is alert and aware of any predators or prey in its direct vicinity. The reticular formation would filter out non-essentials from essentials, allowing the cat to discern what is and what is not important for its survival.

Sensory receptors *

Sensory receptors trigger reactions in your body. Sensory receptors are made up of neurons and other specialized cells. There are several basic types of sensory receptors which all do different things. Several exist in the body of a cat. Thermoreceptors detect heat and cold, mechanoreceptors detect sensations of touch and stretch, pain receptors detect pain, light receptors detect color and movements and chemoreceptors detect taste and smell. The severity of the action which follows a sensory receptor being triggered is a result of strength of the stimulus.

Hearing and Balance *

The ear changes the air pressure wave to action potential which is perceived as sound. Cats have two organs inside their ears, one for hearing and the other for balance. The ear is divided into three regions: Outer ear, Middle ear, and Inner ear. The outer ear is the part we can see or touch. It consists of the pinna and auditory canal, which collect air pressure waves, (sound waves,) and focus it to the eardrum. The eardrum is located in middle ear, and is a sheet of tissue that separates outer and middle ear. When air waves strike the eardrum, the eardrum passes the vibration to the hammer, anvil, and stirrup; three small bones located behind the eardrum. The stirrup is in contact with the oval window, passing the vibration into the inner ear. The middle ear opens a to a eustachian tube, where air can enter to balance the pressure on both sides of the eardrum. The inner ear is filled with fluid-filled channels, cochlea, where the organ of Corti is located within the middle canal. This organ of Corti has an array of hair cells, sensory receptors of the ears, which are embedded in the floor of middle canal. The tip of the hair cells also contact the tectorial membrane. Sensory neurons from the base of hair cells carry an impulse generated by the process of vibration passed on to the hair cells, along with neurotransmitters to the brain through the auditory nerve.

Different parts of the ear function in different ways. Sound waves cause the eardrum to vibrate at a certain frequency. The frequency can be measured in hertz(Hz) or kilohertz(kHz) depending on the number of vibrations per second. As sound waves go through the three bones in the middle ear, the sound pressure gets amplified about 20 times.

The volume and pitch of the sound depends on the amplitude and frequency of the soundwave. The amplitude, height of the pressure, of soundwave gets higher as the volume of the sound gets louder. Furthermore, higher-pitched sounds have greater hertz compared to lower-pitched sounds. the Cochlea can differentiate these sounds because the basilar membrane varies in its length throughout. Therefore, basilar membrane has different sensitivities of vibration at each region. Cats can hear higher pitches than humans do.

Finally, there are certain organs which sense the motion, position, and balance in the ear. Especially, our inner ear's hair cells bend to perceive changes. Semicircular canals detect the movement of the head, and they are arranged to feel movement in all directions; the fluid inside the canals moves and presses the hair cells.

Taste/odor *

The sense of smell and taste are controlled by sensory receptors, which are a combination of neurons and specialized cells which read signals from the outside world and internal organs. Taste and smell are two very similar senses, but they do have their differences. The process of tasting something works in a series of steps. First, the molecules of the ingested substance travel to the taste buds, sensory receptor cells for tasting. There are five different tastes: sweet, sour, bitter, salty, and umami. Umami is the Japanese word for delicious, and represents more savory flavors, as in different meats, for example. It is activated by glutamate, an amino acid.
 In the taste buds, the ingested chemicals bind to membrane embedded proteins called sweet receptors. This process of binding causes a signal transduction pathway, which causes ion channels in the membrane to open. The opening of the channels allows for increased ion flow and a change in receptor potential. The receptor potential is the potential difference in the membrane of the taste bud cell. The stronger the stimulus (taste/flavor), the higher the receptor potential. This receptor potential sends signals into the central nervous system. Back in the taste bud, every receptor cell creates a synapse with an individual sensory neuron. The receptor cell then lets out different neurotransmitters, which cause a flow of action potentials in the sensory neuron of the cell. Those action potentials are what will cause the cat to taste the food it has eaten.

Chemoreceptors are cells that allow the cat to sense taste and smell. Airborne chemicals are detected by chemoreceptors in the nose. Those chemicals are then dissolved in the cat's mucus, which lines the entire nasal cavity. Olfactory sensors are the chemoreceptors located in the upper nasal cavity. They send impulses through axons into the olfactory bulb, located in the brain. Olfactory cells have cilia which stick into the mucus in the nasal cavity. When a chemical is diffused throughout the mucus, these cilia can bind with it. This causes membrane depolarization and creates action potentials. The cat will have a much better sense of smell than any human, a higher concentration of these sensory cells.

Eyes *

To build the eyes of your cat you will need standard pieces of the eye. These include but are not limited to the cornea, iris, pupils, orbits, and retinas. The area which holds the eyeball of the cat is known as the orbit. The orbit also produces and drains tears. The cornea of a cat functions very similar to a human's. It is a spherical shaped area which lets light into the eye and acts as protection for the eye. Another portion of the eye which cats and humans share is the iris. The iris is the part of the eye which is colored and it controls the intake of light. This is done by enlarging or shrinking the pupil. The pupil is the black part in the middle of the eye. When in the dark, the pupil will become bigger to let more light in. In a bright area, the pupil shrinks to let less light in. The tapetum lucidum is a reflective layer which the cat uses to reflect and therefore increase available light which helps them to see at night. It is this which gives the cat a blue or green tint to its eyes at night. Similar to humans, cats have retinas. The retina is a key piece of the eye because it transfers light energy into signals that are then relayed to the brain. When constructing your cat you must remember to include all these parts of the eyes.

Muscle-neuron interaction *

Without nerves and neuron signals, muscles would contract whenever ATP is present, which would be unpleasant for a cat. The cat needs signals from motor neurons to tell muscles when they are needed and when to contract. Motor neurons release a neurotransmitter, which signals an action potential, causeing muscles to contract. The ratio of the nerves that send these neurotransmitters is extremely low compared to the muscle fibers. However, the more used a muscle is, the more motor proteins are activated. When muscles are at rest, regulatory proteins block certain proteins from binding the actin molecules so that they will not contract. They do this by wrapping around a muscle filament when the muscle is at rest and releasing when the muscle is active, or contracting.

Muscle *

A cat can jump, run or turn a circle is because of the interaction between its skeleton and muscles. Therefore, if we want to build an active cat, we are supposed to build many kinds of muscles in its body firstly. They cause its skeleton to move by contraction. However, one muscle can only move the bone in one direction, so we need to build a whole skeletal muscle system of muscle groups to help the cat's body make back-and-forth movement. We also need the connections between muscles and skeletons, which are called tendons.

If we cut the muscle in half, we find it is composed of a hierarchical structure of parallel strands. It means it is made by smaller and smaller units. The biggest unit is muscle fiber, which is a cylindrical cell with nuclei. After this, each muscle fiber is consisted by many strands of myofibrils that contain contractile proteins actin and myosin. Lastly, the fundamental unit of a skeletal muscle is a sarcomere. Series of these make up the myofibril. A sarcomere is composed of thin filaments of actin molecules and thick filaments with myosin molecules. Also, there is another protein called troponin in thin filaments, which mainly regulates the muscle contraction.

The contraction of muscles largely depends on the movement of the structure of a sarcomere. Basically, once thin filaments slide along thick filaments, the muscle contracts. The Myosin molecules provide energy to the movement. They lie parallel on thick filaments, and each of them has a head that sticks out of the filament. The head has two binding sites that can bond to the actin molecule of thin filaments and to ATP, where it hydrolyzes it to release energy for fueling muscle contraction. After binding to ATP, it switches from a low-energy position to a high-energy position. The myosin head then extends toward the actin molecule on thin filaments and bind to it, which causes a connection called cross-bridge. It returns back to its low-energy position, which pulls the thin filament to the center of the sarcomere. This process repeats after another ATP binds to the myosin molecule.

Skeleton *

The vertebrate skeletal maintains the structural support to allow movement in both the cat and mouse. Creatures with a backbone, or spinal column, are vertebrates. All vertebrates have a skull, backbone and almost all vertebrates have a rib cage. These three parts of the skeletal system make up the axial skeleton which holds up the main part of the body. The skull surrounds the brain and provides protection from outside contact. The backbone protects the spinal cord and is made up of individual bones called vertebrae. All vertebrates contain a distinct and different amount of vertebrae. Within most vertebrates, a rib cage encompasses its lungs and heart. Also, most vertebrates have an appendicular skeleton which is a series of bones that form the vertebrate's arms and legs. This part of the skeletal system allows vertebrates to be active and to participate in locomotion. The skeletons within both the mouse and cat contain hundreds of bones that are organs formed by living tissues. Bones comprise the skeletal system and work as a defense barrier. However, joints cause the skeleton of the cat and mouse to become more flexible and engage in movement. Joints occur at intersections between bones that allow separate movement. When building a cat, it is highly important to understand and include all of these parts of a skeleton.

(And second, the apple tree...)

Angiosperms, dicots (apple) *

There are two main types of vascular plants: gymnosperms, (naked seed), flowerless plants, and angiosperms, which are flowering plants. There are 250,000 different species of angiosperms, including orchids, bamboos, and, in this case, apple trees. In the seed of an angiosperm, are two seed leaves, (cotyledons), to make it a dicot ("two cotyledons"). Other characteristics of dicots are branched leaf veins, a ring arrangement of phloem and xylem, or vascular, bundles in the stem, flower petals arranged in multiples of 4 or 5, and a main root, known as a taproot, from which other roots, known as lateral roots, branch.

Basic plant structure *

The plant body is specialized to absorb water and minerals from below ground and CO_2 and oxygen from above the surface. The organs of a plant can be sorted into two systems: roots and shoots. Neither system could survive without the other, as roots would starve without the sugars shoots obtain, and shoots would die without the water and minerals the roots absorb.

The root system holds plants in the soil and, as stated above, obtain and transport water and minerals from the soil. The roots also store carbohydrates. Near the tip of the roots there are outgrowths of epidermal cells (cells in the outer layer of the root) called root hairs. The root hairs increase surface area and allow for the roots to efficiently absorb water and minerals.

The shoot system includes the leaves, stems, and the reproductive structures, flowers. The stems support the leaves. The part of the stem connected to the leaf is called a node. The part of the stem between nodes are called internodes. The leaves themselves are the parts of the plant which mainly perform photosynthesis, though some stems also perform photosynthesis.

The buds of a plant are shoots which have not developed yet. There are two types of bud: the terminal bud, which is at the tip of the stem, and the axillary bud, which is in a "crook" formed by a leaf and the stem. When a plant stem grows, the terminal bud has developing leaves and a compact series of nodes and internodes. In an evolutionary adaptation called apical dominance, the terminal bud produces hormones which stops the axillary bud from growing. This makes the central stem of the plant stronger and taller, which increases the plant's exposure to light. Under certain conditions, such as the removal of the terminal bud, the axillary bud grows. The bud may develop into shoots with flowers or branches with their own leaves and buds.

Primary growth *

Primary growth occurs in the apical meristems, the meristems at the tips of roots and in the buds. Cell division in them produces new cells which lengthens the plant. Primary growth allows roots to push through the soil and shoots to grow. As a result, plants can be exposed to more light.

The growth occurs behind the root tip where the three zones-the zone of cell division, the zone of elongation, and the zone of differentiation, overlap. The zone of cell division produces new root cells and the cells of the root cap. In the zone of elongation, root cells grow longer than their original length. The elongation of the cells pushes the root tip deeper into the soil. Because of the circular arrangement of cellulose fibers in cell walls, the cells lengthen instead of expanding in all directions. In the zone of differentiation, the specialization of cells structure and function occurs because of differential gene expressions.

Secondary growth *

Secondary growth occurs in plants following their primary growth. However, secondary growth only occurs in woody plants, such as trees and shrubs. Secondary growth is the thickening of the stems, roots, or trunks of a specific plant. Secondary growth is caused by the dividing of cells in tissues. These tissues are known as lateral meristems. The vascular cambium is a cylinder of meristem cells which is one cell thick. It is located between the primary phloem and the primary xylem. The vascular cambium is completely responsible for secondary growth.

Secondary growth occurs when layers of vascular tissue are added to the cambium. Tissues created by secondary growth are known as secondary tissues.

Thus, two new tissues are created, called the secondary phloem and... you guessed it, the secondary xylem. The secondary xylem is added to the interior while the secondary phloem is added to the exterior. Secondary xylem will build up the wood of a tree or other woody plant. Over time, the woody stem or trunk will become thicker and thicker, as layers of secondary xylem are built. This creates annual growth rings in the tree. These layers are visible because of uneven activity of the vascular cambium from year to year. In temperate regions, such as the United States, the vascular cambium does not create new tissues during the winter. This is what creates the unevenness of the growth rings, and creates a varying thickness between the rings.

However, what occurs externally is quite different. Everything outside of the vascular cambium is known as bark. As the diameter of the stem increases, the external layers fall off or crack. The new outer layer is called cork, which is built by the cork cambium. The original cork and cork cambium will eventually be forced outward. A new cork cambium is built by the secondary phloem, and thus, a new cork. The secondary phloem, cork cambium, and cork make up the bark.

Other parts of stems include the wood rays and the heartwood. The heartwood is the innermost center of the trunk. It is made up of older secondary xylem which no longer transports water. The xylem is filled with resins and other substances which prevent the heartwood from rotting. This heartwood no longer conducts water. So, a large tree can survive without its heartwood. The wood rays are outside of the heartwood. The wood rays transport water and other nutrients throughout the plant. Also, the wood rays store some nutrients and aid the repair of wounds. Another area of the tree is the sapwood. The sapwood is lighter than the heartwood, and is made up of younger secondary xylem, which move xylem fluid.

Flower *

The flower is the reproductive organ of angiosperms. Angiosperms, as the name implies, possess covered seeds. The seeds must also be in the carpel, which contain the stigma and the ovary. The carpel is the stalk in the center of a flower, and the stigma is the top of the carpel. The carpel holds the ovary. The ovary holds the ovule, which is the developing female egg. This reproductive complex is known as the pistil. The life cycle of an angiosperm starts when the flower is fertilized. Then the ovary begins to develop into a fruit. The ovule will become the seed, which contains the embryo. Under proper conditions the seed germinates, or begins to grow into a seedling, which will then become a mature plant.

Pollen/ovules *

As the plant undergoes its life cycle, the generations, haploid (n) and diploid (2n), alternate. Anthers in the flower produce pollen, which is actually the male of the next generation. Ovules are the females of the next generation. Anthers are attached to the upper part of the filament. Usually, after going through meiosis, which is the cell division process, the cell forms four haploid spores. Each spore then goes through mitosis and products are two haploid cells, which are tube cell and generative cell. The two cells are protected in a thick wall surrounding them. The combination of the two cells are contained within pollen grains, which will be released by the anther. An ovule is made of a central cell surrounded by smaller cells. The central cell encounters with the meiosis and produce four haploid cells, three of which deteriorate. The remaining cell undergoes mitosis to produce an embryo sac. The embryo sac is the female gametophyte. It has a large central cell with two haploid cells on the side; one of the two is an egg cell.

Fertilization requires pollination, which is the transfer of pollen from the anther to the stigma. This job may be done by insects, birds, many other animals, and even wind. The tube cell in the pollen grain develops a pollen tube, sticking straight down to the ovule, while the generative cell forms two sperms by mitosis. Once getting in touch with the ovule, the pollen tube enters the ovary at the bottom of the ovule and inserts the two sperms near the embryo sac. Here, a double fertilization occurs. One sperm fertilizes the egg to form diploid zygote, while the other joins the two central cells to form a triploid nucleus, called endosperm. The endosperm develops into food for the developing young plant.

Seed formation *

The ovule, after fertilization, begins developing into a seed. The formation of a seed begins when a zygote divides into two cells. Such a division doesn't stop here; with more and more new cells, there is a ball of cells now called embryo. The functions of cells from different parts of the seed gradually become clear: there are cotyledon, seed coat, endosperm, shoot and root. Now, as the embryonic development progresses, a mature seed is formed. From here on, the seed becomes dormant until the environmental conditions are favorable for it to germinate.

Fruit formation *

A fruit is a vessel made for protecting seeds and dispersing them. It develops from the plant's ovary, after hormonal changes cause it to mature and grow. Fruit often attracts animals who, upon eating the fruit, serve to disseminate the seeds within.
Transpiration *

Transpiration is the loss of water from a plant's leaves by evaporation. Transpiration pulls water up xylem vessels. Plants need a constant supply of the soil's dissolved minerals and water. If the soil dries and transpiration out of the leaves exceeds the rate of water delivery to the leaves, the plant will wilt and die unless it gets rehydrated. This supply of the soil's dissolved minerals and water is provided as xylem sap, a solution of water and inorganic nutrients. It flows from the roots to the leaves through the shoot system. It flows through extremely thin tubes in xylem tissue pulled by transpiration. The tension created by transpiration works to pull the string of water molecules upwards.

Transpiration is controlled by guard cells. A pair of guard cells control the opening and closing of a stoma by changing shape. The opening and closing of the stomata helps balance the plant's need to conserve water with its requirement for photosynthesis. A stoma opens when its guard cells gain K+ ions and water from neighboring cells. A stoma closes when its guard cells lose K+ ions and water. During the day, the guard cells may close the stoma is the plant is losing water too fast.

Phloem and Xylem *

In the apples, and all angiosperms, whose energy might eventually be consumed by cats, phloem and xylem act as internal transport systems. Phloem is used to transport the products of photosynthesis in a liquid appropriately called phloem sap, while the primary function of xylem is to transport a mixture of water and dissolved minerals, termed xylem sap.

Phloem sap requires perforations in the sieve plates, which connect the cytosol of each cell, to passively transport itself freely among cells. The sap's main solute is the disaccharide sugar, sucrose. The additional contents of the sugary phloem sap includes amino acids, hormones, and inorganic ions.

Xylem sap is transported from the roots to the leaves through extremely thin tubes which are contained the **xylem tissues**. The act of transporting xylem sap is passive meaning it requires no cellular energy to move it up the plant. This is mainly assisted by capillary action, root pressure, and transpiration.

Plant hormones *

Hormones are chemical messengers produced in part of the plant that deliver their message in another part of the plant. Plants have five classes of hormones, auxins, gibberellins, cytokinins, ethylenes, and abscisic acids.

Auxin hormones promote cell growth and cell expansion, and therefore are produced primarily in parts of the plant which are actively growing, such as the very top of the stem. They are transported in one direction, from the top of the plant to the bottom of the plant and is the only plant hormone known to do this. Auxin also maintains apical dominance and prevents many lateral buds and branches from growing on the side of stems. They also move to the shaded side of the plant stem and promoting those cells to grow larger. The cells on the sunny side will stay the same size and will cause the plant to bend to one side.

Gibberellins play an important role in several developmental stages in plant, but are known for promoting stem elongation between nodes of the stem. A node is a place on a stem where a leaf attaches, and the gibberellins elongate the internodes.

Cytokinins are involved in cell division and the making of new plant organs, such as roots. They are produced in the very tip of the roots and travel up the stem through the xylem and is important in plant repair. Cytokinins delay senescence, the natural aging process in plants, which leads to death and encourage cells to divide.

Ethylene is a plant hormone that impacts the ripening and rotting in plants. Ethylene, which exists as a gas, can be produced in almost any part of a plant, and can diffuse through the plant's tissue, outside the plant, and travel through the air to affect other plants.

When the plant needs water it produces a chemical messenger, abscisic acid, to alert the rest of the plant. Abscisic acid is made in droughted leaves, droughted roots, and developing seeds. It can also travel both up and down in a plant stem in the xylem or phloem sounding the alarm. Abscisic acid travels to the guard cells, telling that water is scarce. The guard cells in response move charged particles out of themselves, which subsequently causes water inside the guard cell to leave, too. The guard cells shrivel and the stomata on the bottom of the leaf close, not allowing water to exit the plant through the stomata.

Ethylene *

Ethylene is the reason an apple ripens and falls from the tree. It is a gas produced by plants to speed up the fruits' aging process or to trigger its programmed cell death. This gas can be passed from one apple to another through the air and cause the other fruit to ripen as well. Therefore one fruit can spoil every single fruit in the same box. Ethylene is also the cause of the color change and falling of leaves. It breaks down the chlorophyll in the leaves and the layer between the leaves and the stem of the tree. This helps preventing the tree from drying out in the winter by slowing down its rate of evaporating water.

Each cat must be slightly different from all other cats, so some individuals in a population may survive changes in the environment. Therefore, the instructions for building a cat must have a built-in mechanism that allows for variability.

Mutation and Reproduction Create Variation *

To begin, it is paramount that one understands the significance of genetic variation between organisms in a species. Without this difference, evolution could not take its course. The two primary sources of genetic variation are mutation and reproduction.

Reproduction is an important factor in creating genetic diversity. Reproduction creates genetic variation by introducing new gene combinations in offspring, from the parents. Put simply, genetic material is 'reshuffled', producing offspring with specific genetic material different from that of their parents. To be understand better, consider this: a father has a large nose, and the mother has small ears. The reshuffling of genetic material could make it possible so that an offspring possesses both these traits.

A mutation is a change in DNA. DNA determines who we are –behavior, personality, appearance, body shape, etc. So any changes within it will change just that. Sometimes, relatively small alterations can have a large effect, but evolutionary changes typically occur due to multiple mutations. It is important to note that there are mutations called 'germ line mutations.' These are mutations that take place in reproductive cells, and can be passed to offspring through either egg cells or sperm cells.
It is a common misconception that mutations occur (in an evolutionary sense) to benefit the organism. Mutations actually happen at random, and can either help, make no obvious change, or harm the organism.

Each individual organism has individual phenotypic variations, or differences such as appearance, behavior, psychological characteristics, etc. Mutations can cause changes in phenotype (important alterations, of which can be lethal, or harm the organism on a large scale, such as decreasing resistance to certain diseases), small changes in phenotype (less noticeable changes in the characteristics of an organism, such as heterochromia, a difference of the coloration of eyes, hair or skin) or no change in phenotype (mutations at with no effect in phenotype.)

Natural selection *

Natural selection is the process of "adaption" according to the environmental condition. The idea was first introduced by Darwin, after he observed artificial selection, in which humans domesticate wild animals and, over generations, turn them into animals which are more suitable and helpful to them. An example here is the domestication of wolves, the ancestors of dogs. Humans selected for the desired traits.

The process of natural selection works like this. Think of one single bug species living in the fields full of wheat. The more brightly-colored bugs will be easier locate by the predators, so the wheat-colored bugs will survive and reproduce. This results in more wheat-colored bugs.

There are three essential key points about natural selection. First, it is not the individuals that evolve. It is a shift in the allele frequency in a population over time due to selective, environmental pressures. Second, natural selection only applies to heritable traits, not acquired, because they are not given by the parents. Third, evolution does not have an ultimate goal. Evolution is only considered as "adaptation" to a variable environment. It is not directed toward the creation of a "perfectly adapted organism" which that can survive in all kinds of conditions. In other words, certain traits of a species might be completely useless in a different environmental condition.

The cat, the mouse, and the apple tree are all organisms that interact with each other, and with the abiotic environment.

Population growth (Mice) *

There are two models for measuring population growth: exponential (idealistic) and logistic (realistic). In an exponential growth model, limiting factors such as resources depletion, habitat carrying capacity, and predation are not considered, resulting in an idealistic population growth model where a species will flourish limitlessly without any restrictions. Since it is an exponential model, population growth can be calculated by multiplying the rate of increase by the current size of the population. A more realistic way of examining population growth is the logistic growth model. In this model, population growth is calculated by multiplying the rate of change by the current population size and ratio of the current population of a species to its habitat's maximum carrying capacity. The advantage to this growth model is the function limits the rate of population growth, and thus limiting the population size of a species to the maximum carrying capacity of its habitat.

There are two major types of factors which influence population growth in different species. The first is density-dependent factors. Intraspecific competition is a big part of density-dependent factors. The competition among individuals of the same species for food supply will most likely decrease the birth rate, and increase the death rate of a certain species, causing population fluctuation. The availability of habitable space is also a vital density dependent factor since it determines the availability of hiding spots from the predator. A study of white footed mice proves that population density will limit population growth even when resources are abundant. The high population density delays sexual maturation in the mice population, leading to fewer birth. The second major factor is density-independent factor. Abiotic factors such as environmental influence, human intervention, and natural disasters all contribute to a drastic and irregular change in a species' population. Studies on different species of animals prove that some species maintain a stable population that is relatively close to the maximum carrying capacity, whereas most populations shown fluctuations in long-term data, indicating the mixture of both factors play a role in regulating the population.

Mouse Behavior-genetic/environmental *

If one were to observe a mouse's personality - such as their tendency to be aggressive, or their preferences of food - one could conclude these traits evolved from both genetic and environmental sources. Varying from species to species, all animals have similar and dissimilar traits. These traits are called **phenotypes**, or discernable traits which are developed in an organism from their environment and genetics. A mouse cannot survive in the world with just its body alone. A mouse needs to use its instincts to eat, sleep, and breed with other mice.

Phenotypes fluctuate in mice quite vastly. Genes determining behavior are fairly fixed in a given mouse, but behavior can be altered by environmental factors. The easiest way for the cat to eat this mouse is if the mouse has weak defensive traits. Diet or timidity make a mouse more susceptible to a cat's predation.

Also, should a mouse pup receives less care and attention from its mother, it grows up more likely to become cat food.

Cat - mouse population cycles *

Population cycles are large fluctuations in population which happen regularly, with rapid exponential growth, meaning the growth rate increases quickly, then rapid exponential decay. Mouse population cycles usually span three or four years, where the highest density occurs during the second or third year. The population cycle for domestic cats is from 7 to 10 years without any population control.   The change in population size of the predator effects the prey, and the varying population of the prey effects the predator.

Cat/mouse predation *

In order for mice to avoid their predators, they must adapt through natural selection, (a process in which organisms tend to survive and reproduce more successfully because they have adapted to their environment better). Three main types of adaptations mice have in this regard are color pattern adaptations, appetite adaptations, and reproductive adaptations. When mice camouflage with their surroundings, they have achieved color pattern adaptation. This makes it more difficult for the cat to hunt them. Because they have frequent habitat changes, mice need to adapt to the food limitations or changes which that occur in new areas, an example of appetite adaptation. Because mice in the wild typically do not live for long periods of time, they must change their reproduction patterns in order to continue producing offspring. Female mice can begin reproducing at almost two months old and can produce about ten litters during their lifetime. Our cat must also adapt in order to maintain a stable food source.

Nitrogen cycle: apple tree, mouse, cat *

Nitrogen is key to many cellular components, such as both nucleic acids and the proteins, so this cat of ours is gonna need an ample supply. Fortunately, 78% of the atmosphere is nitrogen in its pure, basic molecular ($N2$) form. Unfortunately, cats and mice can't do anything with nitrogen in this form. When the animal inhales, it takes in the N2 gas. Then the cat, or mouse breathes out the same percentage. In order for the nitrogen to get to our feline friend, it has to take a more convoluted route. The first stop on our journey is a group of bacterial species inhabiting the soil and the roots of certain types of plants (most notably legumes). These bacteria can perform nitrogen fixation, in which they take this free nitrogen in the pores of the soil and produce ammonium (NH4). Some of this Ammonium is taken up directly by plants, but the rest is nitrified by nitrifying bacteria and is then assimilated by a plant. Then some small creature, say, a mouse, will come along, eat some form of plant matter, say, an apple that fell from an apple tree, and get some nitrogen. Lastly, along comes our cat, who eats this mouse and gets the nitrogen it needs in a form it can use. When any plant or animal matter dies, decomposers will break it down and release, among many other things, the nitrogen contained therein, which is released either into the soil or the air and the cycle continues. In order for our cat to tap into this cycle and get the nitrogen it requires, all you need to do is crack open a can of cat food. Everything that the cat eats was, at one point, composed of living cells - mostly meat, but also some plant matter, or milk. Understanding the role our cat plays in the nitrogen cycle requires no advanced knowledge of biology. If you've previously owned a cat, you've already done it before.

Energy flow/chemical cycling *

Energy flow is the passing of the energy through different parts of the ecosystem. Sunlight shines on a plant, which is a producer. The plant then converts the sunlight into chemical energy through photosynthesis and stores the energy inside itself. By chance, a plant-eating mouse consumes the plant. By eating the plant, light energy which was changed to chemical energy by the plant is transferred to the animal's body. When the animal dies, decomposers utilize the energy taken in by the animals. This transfer of energy is not infinite. For every transfer of chemical energy there is some energy loss in the form of heat. The sun drives the energy flow of the whole ecosystem.

Chemical cycling, on the other hand, involves the passing of chemical components in the ecosystem such as carbon and nitrogen. The plants take in these elements through the air and the soil while animals consume those elements from the food they take in. After the death of the plants and animals, the decomposers recycle the elements back into the ground and the air in its inorganic form. Animals also return these elements to the system through feces, urine and skin-shedding.

Carbon cycle *

Although the amount of carbon is finite and fixed, it is dynamic, moving between living and nonliving things. This is the carbon cycle. Carbon cycles through the biosphere, moving from living to non-living and back again.

The Biosphere consists of all carbon in land-living organisms, along with the carbon in soils. There are roughly 2,000 gigatons of carbon stored in both these soil organisms and above-ground organisms.
The carbon in this system can leave an ecosystem through rivers, streams, erosion, as well as through combustion and respiration which release carbon into the atmosphere. Cats and mice exhale carbon as a product of carbohydrate breakdown, which provides them with the sun's trapped energy. The apple tree traps the sun's energy, and locks it into the chemical bonds of the apple.